Applied Sedimentology

Applied Sedimentology

Prof. R.K. Sukhtankar
M.Sc., Ph.D., F.G.S. (Ind.), F.M.A.Sc.
Formerly Prof. & Head, Department of Geology,
Shivaji University, Centre for P.G. Studies,
Solapur - 413 003

CBS

CBS PUBLISHERS & DISTRIBUTORS PVT. LTD.
New Delhi • Bengaluru • Chennai • Kochi • Kolkata • Mumbai • Pune

ISBN: 81-239-1052-5

First Edition: 2004
Reprint: 2008, 2010, 2016

Published by:
Satish Kumar Jain for CBS Publishers & Distributors Pvt. Ltd.,
4819/XI Prahlad Street, 24 Ansari Road, Daryaganj, New Delhi - 110002
delhi@cbspd.com, cbspubs@airtelmail.in • www.cbspd.com
Ph.: 23289259, 23266861, 23266867 • Fax: 011-23243014

Corporate Office: 204 FIE, Industrial Area, Patparganj, Delhi - 110 092
Ph: 49344934 • Fax: 011-49344935
E-mail: publishing@cbspd.com • publicity@cbspd.com

Branches:

- ***Bengaluru:*** 2975, 17th Cross, K.R. Road, Bansankari 2nd Stage, Bengaluru - 70 • Ph: +91-80-26771678/79 • Fax: +91-80-26771680 E-mail: cbsbng@gmail.com, bangalore@cbspd.com
- ***Chennai:*** No. 7, Subbaraya Street, Shenoy Nagar, Chennai - 600030 Ph: +91-44-26681266, 26680620 • Fax: +91-44-42032115 E-mail: chennai@cbspd.com
- ***Kochi:*** Ashana House, 39/1904, A.M. Thomas Road, Valanjambalam, Ernakulum, Kochi • Ph: +91-484-4059061-65 Fax: +91-484-4059065 • E-mail: cochin@cbspd.com
- ***Kolkata:*** 6-B, Ground Floor, Rameshwar Shaw Road, Kolkata - 700014 Ph: +91-33-22891126/7/8 • E-mail: kolkata@cbspd.com
- ***Mumbai:*** 83-C, Dr. E. Moses Road, Worli, Mumbai - 400018 Ph: +91-9833017933, 022-24902340/41 • E-mail: mumbai@cbspd.com
- ***Pune:*** Bhuruk Prestige, Sr. No. 52/12/2+1+3/2, Narhe, Haveli (Near Katraj-Dehu Road Bypass), Pune - 411041 Ph: +91-20-64704058/59, 32342277 • E-mail: pune@cbspd.com

Representatives:

- Hyderabad: 0-9885175004
- Patna: 0-9334159340
- Nagpur: 0-9021734563
- Vijayawada: 0-9000660880

Printed at:
J.S. Offset Printers, Delhi

PREFACE

Preface

The University Grants Commission has a programme of 'Preparation of the University Level Books' The present book entitled, 'Applied Sedimentology' is an outcome of the same. Considering readers of the book as students at the postgraduate and also at the undergraduate level, who offer Geology as a major subject, the book has been written in simple and explicit language. While elaborating on the subject of sedimentology, applied aspects have been emphasized, without ignoring it as primary and fundamental science.

Development and advancements in petroleum geology, advent of computer application, and now the established concept of plate tectonics in earth science, have widened the frontiers of research, particularly in the field of sedimentology. It is a difficult task to keep track of all such recent researches. Number of foreign journals and books are available which attempt to incorporate recent developments in sedimentology. Such books are related to various themes, like sedimentary rocks, sedimentology, sedimentary environments, diagenesis or specific rock types, such as, sandstones, carbonate rocks, etc.. Considering the length, breadth and depth of the subject, it is rather difficult for any book to include all the aspects.

The present book has been written to incorporate new concepts, new ideas in sedimentology to the maximum extent possible, and at the same time, without any disregard to basic, fundamental and therefore, old concepts and theories. It has been noted that on the old concepts, new premises in sedimentology have been built. It is hoped the present book should cater to the needs of geology students from any Indian University and even to those from university of any developing country. It is based on most of the postgraduate syllabi and is also according to the model syllabus prepared by Curriculum Development Centre in Earth Science, formulated by the University Grants Commission.

Accordingly, text of the book has been spread through fourteen chapters. Starting with origin and formation of sediments, processes responsible for their origin has been followed by their textural attributes and methods of analysis, presentation and computation of textural data and their significance and interpretation in sedimentological research. Different types of syn-depositional and post-depositional sedimentary structures have been described alongwith their characteristics, origin and other relevant aspects. Various attempts and standard schemes of classification of sedimentary rocks and facies concept and depositional sedimentary environments have also been dealt with fairly in detail. Of the different lithologies, various aspects of major rock types, viz; sandstones and carbonate rocks have been described in detail, while only significant aspects of minor rock types have been described.

It is hoped that the book, 'Applied Sedimentology' will serve the purpose of students, researchers and teachers in Geology from the point of view of study and research.

R. K. SUKHTANKAR

ACKNOWLEDGMENT

Acknowledgment

I am much obliged to the University Grants Commission, especially the Chairman and the experts from the Earth Science panel, who bestowed upon me a great responsibility of writing the book, entitled, 'Applied Sedimentology'. Let me express, therefore, my sincere and grateful thanks to U.G.C. -Earth Science panel for identifying me in writing the University Level Book. I am afraid, and really wonder, if without their patronage and encouragement I could have ever engaged myself in such a task. Once given the responsibility, an attempt has been made in writing this book, to comply with the requirements and expectations of all post-graduate students and teachers of Indian Universities.

I find no words to express my feelings for sincere, painstaking and unreserved help extended by my colleagues, Dr. D. D. Kulkarni and Mr. M. A. Kamble. Both of them helped me at every stage of the progress of the book. Without their assistance to complete the book and to get it in the final form would have been a difficult task. My honest thanks to both of them.

Constant encouragement and moral support received from Dr. S.K. Guha, Prof. Emeritus, Department of Space Physics, University of Pune, and Prof. K.B. Powar, Secretary General, Association of Indian Universities, New Delhi is gratefully acknowledged with thanks. I am indebted to Prof. G.G. Deshpande (retd.), Department of Geology, Nagpur University, Nagpur, for introducing and cultivating in me an insight into the field of Sedimentology, which has helped me in writing this book.

My friends and acquaintances from the Geology Departments of different Universities and research institutes and organisations have helped me in the preparation of this book by way of providing either relevant literature or their research publications. Particular mention can be made of Prof. T. C. Devaraju, (retd.), Prof. V. C. Chavadi and Prof. M. S. Mannikeri of the Department of Geology, Karnataka University, Dharwad; Dr. N. H. Hashimi, Dr. Rajiv Nigam, Dr. O. S. Chauhan, Dr. B. G. Wagle, Dr. A. R. Gujar of the Geological Oceanography Division, National Institute of Oceanography, Goa; Prof. G. V. Rajamanickam, Department of Geology, Tamil University, Thanjavur; Prof. K. R.Subrahmanya and Prof. R. Shankar, Department of Marine Geology, Manglore University, Mangalgangotri; Prof. A. G. Dessai, Dr. Mahendra and Dr. P. S. Raikar, Department of Geology, Goa University, Goa; Prof. N. M. Vashi (retd.), Prof. M. P. Patel, (retd.), Prof. Nikhil Desai, Prof. S. Ganapathi, Prof. L. S. Chamyal, Dr. A. V. Joshi and Dr. B. Deota of the Department of Geology, M. S.University, Vadodara; Dr. P. M. Tapaswi and Dr. A. G. Bhusari, Department of Geology, Nagpur University; Dr. Neloy Khare. National Centre of Antarctic Ocean Research, Goa and Dr. C. Rajshekhar and Dr. V. D. Borkar, Department of Geology and Palaeontology, Agharkar Research Institute, Pune. I am thankful to one and all for their kind gesture.

Cooperation and help received from Prof. N. N. Maldar, Head, Department of Chemistry, Shivaji University Centre, Solapur; Dr. P. T. Sawant and Dr. S. J. Pandit, Department of Geology, Walchand College, Solapur, Dr. S. A. Kulkarni, G. S. D. A., Government of Maharashtra, Dr. A. J. Shirke, Department of Geology, College of Engineering, R. I. T., Islampur, and Mr. R. N. Jadhav, Co-ordinator, AIHC & Archaeology, Shivaji University Centre, Solapur, at various stages during the preparation of this manuscript is gratefully acknowledged with thanks.

Members of the teaching and non-teaching faculty of the Department of Geology, Shivaji University Centre, Solapur, helped me at various stages of the preparation of the manuscript of this book. In particular, Dr. P. D. Mali helped me in photography and Mr. D. T. Gujar helped in drafting some figures. Help and assistance rendered by everyone is acknowledged with thanks.

The authorities of Shivaji University and its administrative staff, mainly from STAT Section, under the Director, Board of College and University Department, Kolhapur were kind enough to release the funds, as and when required during the progress of writing and completing the manuscript. thanks are also due to them.

I would be failing in my duties, if I do not express my grateful thanks to the U.G.C. authorities, who were considerate and kind enough in granting me an extension in time limit, initially defined for the project. The extension was to be sought due to physical constraints I had to face, to which U.G.C. authorities were sympathetic.

lastly. I express my indebtedness to my family members, to my parents, my wife Padmaja, son Advait and daughter Prajakta, who held ungrudging attitude towards my long absence from home.

R. K. SUKHTANKAR

LIST OF FIGURES

List of Figures

CONTENTS

Contents

1 CHAPTER

Introduction

1.1 INTRODUCTION

Sedimentology is the science of sediments. It is a significant and one of the important core branches of Geology. Nearly 66% of the continental rocks are either sedimentary or have been formed from the earlier sedimentary rocks, while entire oceanic surface is covered by sediments. Apart from hosting certain economic minerals, natural resources of energy, like petroleum, natural gas and coal, occur in sedimentary rocks. Exploration and exploitation of such natural resources have led to the development of the subject of sedimentology, not only from the academic but also from the applied aspects.

In earth science, three different terms that relate to the study of sediments, are commonly encountered. However, each has a different connotation and deals with different aspects of scientific study of sediments. These three terms are sedimentology, sedimentary petrology and stratigraphy.

The term 'sedimentology' has been defined by Wadell (1932) as 'the study of sediments'. In Chamber's dictionary (1972), sediments have been defined as, 'what settles at the bottom of a liquid: a deposit'. Sedimentology also includes chemical precipitates, like salt, as true detrital deposits. Boundaries of sedimentology, therefore, appear to be diffused.

Sedimentology essentially includes study of natural processes such as; erosion and weathering, responsible for origin and types of sediments. Another major aspect deals with the study of agents, responsible for transportation of sediments. Thus, sedimentology essentially includes the scientific study of sedimentary processes, but excludes the subject of diagenesis. Apart from dealing with the nature, types and classification of sedimentary rocks, sedimentary petrology includes study of processes that operate in the realm of environments of deposition. These are compaction, lithification and diagenesis, and which are post-depositional, i.e. after deposition of sediments has taken place. These processes describe various physical, chemical and mineralogical changes that are responsible for conversion of sediments into sedimentary rocks.

Subject of stratigraphy has historical aspect of the study of sediments. It is a study of stratum or strata. Deposition of sediments results into succession of strata. It is a deposition of bed one above another. Thus, stratigraphy is a study of history of sediments and sedimentation, which gets revealed bed after bed. In a given simple stratigaphic set-up, lowermost bed encountered in a succession or in a vertical section is considered as the oldest and the first to be deposited in that succession. Deposition of succeeding, overlying beds has resulted one after another. Each bed, therefore, is a repository of events, thereby documenting history of sedimentation. Stratigraphy, therefore, is a historical study of sediments; relating to palaeogeography and palaeoecology. Sedimentary tectonics is another aspect that deals with the relation of tectonics in formation of sedimentary strata. It is the scientific study of tectonic aspects, responsible for deposition of the sequence.

1.2 SEDIMENTOLOGY AND OTHER SCIENTIFIC DISCIPLINES

The word sediment is derived from Latin, 'sedimentum', which means "a settling", implying thereby that settling occurred through air or water. The term 'sediment' refers to regolith that has undergone transportation. Sediments are derived from the earlier, pre-existing rocks by processes of weathering that brings out disintegration and decomposition of the rocks.

In disintegration, rock breaks into fragments of various sizes. Such fragments are transported either under the action of gravity or by another agent of transport, such as, air, water, ice. Deposition of such loose fragments at the foot- hill is termed as colluvium. Disintegration or fragmentation of rocks results due to differential heating, which develops fine fractures and cracks. Weakening and loosening of rocks is facilitated by already existing joints and fractures. Percolating waters through them further brings out decomposition of rocks, thereby accelerating fragmentation. In high latitude countries, where cold climate prevails, thawing action exerts tremendous pressure through fractures and joints, resulting into fragmentation. In hilly or mountainous region, landslide is a common feature, generally in a rainy season. Even earthquake tremors of small magnitude can dislodge huge rock-masses, resulting into landslides. The phenomenon is referred to as mass wasting. Such loose, fragmented rock material is further transported down the slope through streams and rivers, that gradually brings out reduction in their shape and size. uring this process, however, original mineralogical and chemical composition of rocks do not change.

Origin of sediments by decomposition is essentially a chemical process, referred to as chemical weathering. Extent of decomposition depends on mineralogical composition of rocks. Chemical parameters, such as, Eh-pH of the fluid, more commonly water, and prevailing temperature conditions decipher nature and degree of chemical weathering of rocks. In this process, mineralogical stability is of prime importance. Order of mineralogical stability in the process of chemical weathering is same as minerals of the continuous and discontinuous series of the Bowen's Reaction Series from Igneous Petrolgy. Role of chemistry, therefore, is of fundamental nature in origin of sediments.

After sediments are formed, they are transported by different agents, major modes of transport being, by suspension, traction i.e. transport along the bed and saltation, which is intermittent suspension and intermittent traction. The way sediments are transported depends on various factors; such as, density/specific gravity of fluid and also of the sediment particle, velocity of fluid, slope of the channel through which sediments are transported, size and shape of the sediments, etc. Thus, physical and chemical properties of sediments and hydrological characteristics of the fluid decipher mode of sediment transport.

Chemical weathering brings various elements from the source rock in solution. Such elements in a dissolved state give rise to chemical sediments under favourable environmental conditions, which may be supersaturation of a particular element by precipitation or by evaporation in a fluid. Limestone, dolomite, phosphorite, and gypsum are some such examples of the chemical sediments. Laws of thermodynamics govern the formation of chemical sediments, and thereby the chemical rocks.

Sedimentary rocks are potential reservoirs of energy resources, such as; coal, petroleum and natural gas. Study of their genesis, occurrence and methods of exploration is included under the title of 'Fuel Geoloy'. Exploration methods are based on the principles of geophysics. Commonly followed methods are electrical prospecting, seismic prospecting and the like.

It is, therefore, clear that sedimentology is, not only related but rather is interrelated with subjects, like hydrology, fluid dynamics, economic geology, mineralogy, coal and petroleum geology; in addition to the fundamental subjects like physics and chemistry.

Biology is another subject that is also related to sedimentology, as remains of plants and animals, get preserved in sediments in the form of fossils or impressions. These studies are carried out under the heading, 'Palaeontology'. It is a study of fossils and encompasses taxonomy, habitat, the environment in which a particular fossil species had lived. These aspects reveal palaeoenvironments of deposition and palaeoclimatological conditions responsible for formation and deposition of sedimentary strata.

Various studies and research in sedimentology yield a variety of data on different scales. Such data is analysed by statistical methods, such as; univariate analysis, bivariate analysis, multivariate analysis and also with the help of various computer software programmes. Basin studies and modelling sedimentary basins for search of economic minerals, petroleum, etc. are also the advance analytical methods.

Thus, the scientific disciplines, such as; physics, chemistry and biology, and statistics and computer science, have now become an integrated part of sedimentology.

1.3 SEDIMENTOLGY AND THE EARTH SCIENCE

As stated, sedimentology is one of the core branches of Geology, and therefore, it is closely related to other branches of Geology. The world's prime energy resources, viz., petroleum, natural gas and coal are hosted mainly by sedimentary rocks. Exploration and exploitation of these resources are responsible in development of sedimentology. Study of such relationship of energy resources and sedimentology is a part of economic geology.

Like sedimentology, mineralogy is another core branch of Geology, in which light and heavy minerals are studied to interpret nature of source rocks and of the depositional environment. The concept of mineral stability in the context of chemical weathering states about the sequence in which minerals from the parent/host rock could weather to give rise to sediments.

The subject of stratigraphy is primarily based on sedimentary strata. Identification and description of physical properties of strata constitute the subject matter of lithostratigraphy. Lithostratigraphic correlation of strata in a region and across is carried out on the basis of physical, lithological characters. For the purpose, however, the basic, fundamental unit is sedimentary stratum or laminae. Biostratigraphic and chronostratigraphic correlation are other methods of correlation in stratigraphy; in which similarity of biota in a sedimentary unit and absolute age determination on the basis of radioisotope analysis respectively, constitute major aspects of sedimentary strata. Apart from the continental sediments, studies of sediments from marine, oceanic environment are being carried out to know about physico-chemical parameters and environmental conditions, responsible for their formation and also their economic significance. These studies are the subject matter of marine geology and oceanography.

1.4 SEDIMENTARY STRATA AND THE GEOLOGIC TIME SCALE

As stated earlier, sediments as products of weathering get transported and later deposited in the form of strata. Each sedimentary particle is a derivative of the earlier, pre-existing rock that undergoes weathering. These rocks are source rocks for the sediments, which get deposited in a particular environment of deposition, while preserving field, mineralogical and genetic characteristics of the source rocks or provenance, i.e. the region from which sediments are derived. Imprints of sedimentary processes, such as, aeolian, fluvial, etc. are also preserved in sedimentary strata. In a depositional environment, therefore, a stratum after stratum is deposited one above another, resulting into a succession of sedimentary strata. Thickness of succession differs from place to place, which depends on a number of factors; such as; time and duration of sedimentary process, depth of the sedimentary basin, apart from tectonics that controls the processes of erosion, transportation, and deposition.

Thus, according to the order of superposition, lowermost stratum in the succession is regarded as the oldest, while succeeding strata are younger and younger in age. In any sedimentary terrain, the prime task therefore is to establish the succession of stratified, sedimentary rocks. Geological events as recorded in strata are to be placed in their correct sequence. It was James Hutton (1726-1797), who first indicated the rocks as records of geologic events, with long geologic time span. With this concept, principles of stratigraphy were gradually estblished, in the form of lithostratigraphy, biostratigraphy and chronostratigraphy. Though each of these has based on different principles, interrelationship amongst them needs to be studied and analysed, for which therefore, stratum is a fundamental unit.

Analysis of sedimentary strata in a section yields geologic column at the place. Such sections from different places are correlated thereby to establish regional geologic or stratigraphic column, which with the help of radiometric methods has yielded geologic timescale.

1.5 APPLIED SEDIMENTOLOGY

Of the three different branches of petrology, much advancement by way of research and development has been made in the field of sedimentology. It is mainly because of exploration and exploitation of natural energy resources, as these are hosted mainly by sedimentary rocks. Necessity of these reserves to mankind is responsible for detailed studies and development. In sedimentology, the processes that are responsible for origin of sediments, their transport and deposition are studied to interpret environment of deposition, provenance and its control on nature and type of sediments, various primary and secondary sedimentary structures, fluid dynamics and various textural parameters of sedimentary particles. Yet, sedimentology has much potential in applied aspects, which is due to textural parameters of secondary nature, viz., porosity and permeability. These properties have led sedimentary rocks, as good reserves for groundwater and petroleum and which therefore serve as good aquifers. Hydrogeology and petroleum geology not only know the extent of reserves of fluid in the host sedimentary rocks but also establish the economic status, to plan the further extraction programme.

Coal is another natural energy resource that also is hosted by sedimentary rocks. Coal geology in itself is another developed branch. It includes the study of coal petrography to infer quality of coal, apart from its mining aspects. Sedimentary strata hosting such reserves yield the information about palaeoclimate, palaeoecology, environments of deposition and also the tectonics.

Evaporites, such as; halite and gypsum are economically significant and important deposits, associated with sedimentary rocks. Association of evaporites with the petroleum reserves has increased the importance of their study.

Deposition of sediments at the construction sites of off-shore engineering structures, such as, jetties, harbours, etc. poses problems in their construction and therefore, pattern and way of sedimentation in littoral and neritic zones need to be investigated. Such investigations document slope of the off-shore, tidal behaviour and tidal dynamics to demarcate areas of erosion and deposition. Construction of piers either in jetties or in habours, also drastically affects pattern of sedimentation. Stability of piers, either in jetties or in harbours, therefore, is an important consideration.

Thus, sedimentology plays a significant role in the fields of fuel geology, engineering geology, stratigraphy and other related subjects, emphasising its applied nature, apart from the basic, fundamental academic aspects. Sedimentology, therefore, is of multi-disciplinary nature.

2

CHAPTER

Field Methods

2.1 INTRODUCTION

Geology, as an Earth Science is essentially a field subject. Its studies primarily begin from the rocks that are exposed on the surface in the field. It is therefore of prime importance that field studies are undertaken with an intention to get knowledge in detail about various lithological units exposed in the field. For the purpose, various field methods are followed and a geological map of that region is prepared. During the field studies, nature of contacts of different lithological units alongwith their attitude are noted. Various structural elements, such as; dip and strike, joints and joint pattern, shear zones, etc. are also recorded. During field investigations, representative samples of lithological units are collected for different laboratory investigations. Results of such investigations of sample analysis are then interpreted for lithological units from which the samples have been collected. Methods of field studies to be followed are, therefore, of utmost importance in geology.

2.2 IMPORTANCE OF FIELD WORK IN SEDIMENTOLOGY

Sedimentology primarily needs field investigations of various lithological units, exposed in the field. Methods to be followed in the field are important and significant mainly from the interpretative point of view, as sedimentology bears close relation with sedimentary petrology, stratigraphy, sedimentary tectonics and coal and petroleum geology.

The field data are of help in construction of palaeocurrent directions, basin configuration, palaeoenvironments of deposition, and thereby to construct history of sedimentation. Thus, aprt from an academic aspect, sedimentology has economic significance.

Considering the realm of sedimentology, necessity of knowing nature, attitude and occurrence of sedimentary rocks in the field are of prime importance. The primary requirement, therefore, is selecting a suitable field method, which depends on the aim and objectives of study. After the objectives are defined, it is to map lithology and lithological variations alongwith the attitude of each lithounit, by following a suitable method. Geological map of the region is then prepared. Such a map should indicate apart from lithology, areas of economic significance, marked by the presence of workable ores and ore minerals.

Lithological map is of utmost importance, as one can learn lithological variations and attitude of lithology in the field, and also primary and secondary sedimentary structures alongwith variations, if any. The field information can be documented by following a suitable field method. Primary aim of such a method is to cover the entire areas so selected, a mapping all the exposed lithounits. Generally different methods are followed to prepare detailed geological map.

2.3 METHODS OF FIELD WORK

A field method to be selected should cover entire area under investigation. An initial approach therefore is to carry out reconnoitory survey, with the help of topographic sheets. In such a survey, variations in lithology, nature of contacts between different lithounits, nature, size and shape of outcrops and their attitude in the field are noted.

After the reconnoitory field survey, detailed field studies of each outcrop are undertaken. in which all the lithological properties, as can be observed in the outcrop, and also the textural and structural properties of every litho-unit are noted. Variations in lithology and dip and strike of each lithounit are recorded. After noting these aspects of every litho-unit, preparation of geological map is carried out by traverse mapping, followed by outcrop mapping. As stated earlier, these studies are carried out with the help of topographic maps.

During traverse mapping, variations in lithology, nature of contacts between different litho-units, and attitude of each lithounit in the field are recorded in the note book and accordingly are marked on the topographic sheet or on the overlay. Traverse mapping is generally undertaken by selecting traverses across the strike of the formations, as difference and variations in lithology are encountered.

Traverse mapping is followed by outcrop mapping, in which individual outcrop is surveyed for its size, shape and lithological and structural characters. By noting these properties, each individual outcrop is reproduced on the topo-map or on the enlarged overlay, gradually to complete geological map of the area. Thus, by following the methods of traverse and outcrop mapping, geological map of the area is prepared.

2.4 DESCRIPTION OF OUTCROP SECTIONS

During the field visit, the first that is encountered is an outcrop. It is therefore, important that various lithological and structural characters of an outcrop should be recorded in detail. Such a record, however, depends on field experience and subject knowledge of an observer. It has to be in detail, in less number of field visits, as available time and money impose constraints to repeated field visits. It is, therefore, desirable to collect maximum field information in minimal field visits, to avoid excess on time and money.

To have a complete description of an outcrop, knowledge of origin of sedimentary rocks is essential. Complete description in the first visit may not be possible. After an initial visit, one has to come back to the laboratory and obtain the information by analysing the data. Such an analysis helps to know the missing links or gaps in the field information. In subsequent field visits, it is to observe the same outcrops, lithology and describe them in detail to fill in the gaps encountered earlier. Thus, in subsequent field visits full and complete lithological description of the outcrops can be made. Such a complete description in all aspects of field characters is desirable, as it throws significant light when collaborated with laboratory studies, not only on the genesis of sedimentary rocks but also on evolution of the basin of deposition, in which the sedimentary rocks have formed.

For getting a full and complete description of an outcrop, comprehensive logs, including various aspects of the field characters of an outcrop have been prepared. Such comprehensive logs are valuable aids to the field geologists.

It was Bouma (1962) who has produced the graphical logging technique for recording various field aspects of the flysch sequence of rocks, i.e. mainly sandstone/shale sequence. Such a graphical log has separate columns, each for rock type, bedding type, bedding plane properties, grain size, carbonate content, etc. All the properties indicated in such graphical logs, however, cannot be determined in the field. The same can be revised only after estimation.

Considering the limited applicability of such logs for the graphic display of field data, Walker (1967) proposed a data form to record "flysch" sequence. It is based on the idealised sequence of sedimentary structures in turbidites, which have been recognised by Bouma. The data form contains thickness of beds and subdivisions of idealised turbidite bed, i.e. massive division, plane laminated division and cross-laminated division. With the help of such data form, Walker could complete descriptions of number of such beds in many stratigraphic sections.

Such graphical logs, however, have their own advantages and disadvantages. A major advantage is that the observer attempts to collect all information pertaining to the points, included in the logs. Nothing is left to the memory, as to what the field aspects/characters of an outcrop are to be recorded and a record is made to have a full and complete description of an outcrop, as far as possible.

Such graphical logs have one disadvantage that it is not applicable or it has limitations in the area, where other major rock types are exposed. Similarly, any other field aspect present, if any, but not included in the graphical log is also likely to be missed, about which the observer is unaware of. This is another disadvantage.

Important benefit that lies in such graphical logs is that all the quantitative information for the area studied is available at hand immediately. The same can be transferred to the map and also for analysis in the laboratory.

2.5 METHODS OF SAMPLING

Due to limitation of time and financial sources, limited exposures and accessibility in the field, field geologist is able to examine and collect the information only of a small part of the sedimentary terrain. Due to such limitations in the field, he has to collect such samples for laboratory studies, that are representative of lithological units, exposed in the area. In the laboratory, samples are analysed for various petrographical, mineralogical and chemical characters and data pertaining to each is collected and documented.

Data analysis is further carried out to understand genesis of rock type, evolution of the sedimentary basin and other stratigraphic aspects. In order to be truly representative of an entire sedimentary terrain, considerable importance is therefore attached to technique of sampling.

The basic concept of sampling, therefore, is of a random sample, the concept that has been borrowed in sedimentology from the principles of statistics. By definition, random sample consists of a small number of observations, which are selected in such a manner that these are generally representative of a large number that basically constitutes a litho-unit,being investigated and examined. Such a litho-unit, in terms of statistics, is referred to as 'population' and the random sample is considered as the one in which every member of the population has equal chance of getting represented. From the study of such random samples, inferences are drawn about the population. Thus, the inference is from sample to population and, therefore, the entire theory of drawing valid inferences is based on the assumption that sampling is random sampling.

Before proceeding for sampling in a terrain, a geologist should have the knowledge of lithology and stratigraphy of the area. He should accordingly subdivide the population into different groups or subgroups. Such a subdivision can be geographic areas, where different litho units are exposed or according to stratigraphic principles, as member or formation, etc. A random sample is to be collected from each subgroup, so defined. The whole sample is then referred to as stratified random sample. In such a method of sampling, therefore, the sample is random within each subgroup but systematic within the entire population.

Sampling of the subgroup or formation or a member is accomplished by following various methods, which are commonly the grid methods of sampling. After th nature and attitude of lithounit are established, sampling of the same is undertaken.

Sampling is carried out by one of the following methods. Selection of the method is necessary; as samples collected from the field, are analysed and examined in laboratory for various petrographic studies and results of such analyses are interpreted for the entire lithological unit or units for which the investigations have been carried out. Following are commonly practised methods of sampling.

Grid method of sampling: There are two ways by which grids are selected, viz., square grid and rectangular grid (Fig. 2.1 *a* & *b*). Selection of the type of grid depends upon the nature of an outcrop.

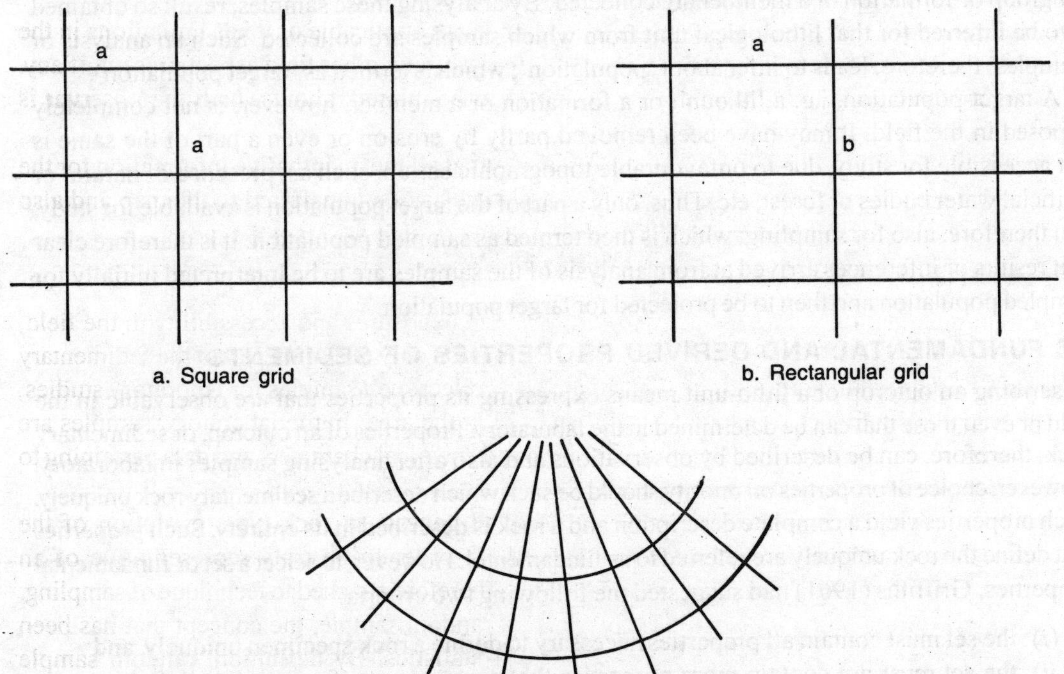

a. Square grid

b. Rectangular grid

c. Radial grid

Fig. 2.1

In a grid method of sampling, grid lines are parallel to each other. To one parallel set of grid lines, another set again of parallel grid lines, cross cuts the earlier set at right angles. Accordingly, it is either a square grid pattern or a rectangular grid pattern. Distance between successive grid lines is based on the nature and extent of outcrop, area to be sampled and purpose of study. A grid pattern is initially drawn on overlay sheet, and is then placed on toposheet or geological map of the area. Each cross point of a grid is a place from where sample is to be collected. In a terrain, where the strata is dipping, one set of grid lines is placed preferably parallel to the strike direction. Another set, being perpendicular to it, therefore, is parallel to the dip direction of the formation. Thus, the sampling is accomplished in both, the strike and dip directions.

However, such a grid method of sampling is not applicable in fluvial, and especially in a deltaic terrain. In such case, a radial method of sampling (Fig. 2.1*c*) is followed. Selection of a common or centre point is based on the nature of delta. In this case, radial lines are drawn according to the pattern of tributories. Thus, on the overlay sheet, the radial lines from the common or central point are drawn, and the constant distance is kept between consecutive arc lines. Intersection points of radial and arc lines are the sites of collection of samples. Such an overlay sheet is then placed on the topo map, sample location points are marked; samples are then collected from the field for further laboratory analysis.

Target Population and Sampled Population

By following a suitable method of sampling, random samples of a particular lithological unit, a subgroup or formation or a member are collected. By analysing these samples, result so obtained is to be inferred for that lithological unit from which samples are collected. Such an analysis of 'samples' therefore, leads to infer about 'population'; which is termed as 'target population'.

A target population, i.e. a lithounit or a formation or a member, however, is not completely exposed in the field. It may have been removed partly by erosion or even a part of the same is not accessible for study, due to unfavourable topographic set-up, such as; presence of natural or artificial water bodies or forest, etc. Thus, only a part of the target population is available for study, and therefore, also for sampling; which is then termed as sampled population. It is therefore clear that results or inferences arrived at from analysis of the samples are to be interpreted initially for sampled population and then to be projected for target population.

2.6 FUNDAMENTAL AND DERIVED PROPERTIES OF SEDIMENTS

Describing an outcrop or a litho-unit means expressing its properties that are observable in the field or even those that can be determined in the laboratory. Properties of an outcrop, or sedimentary rock, therefore, can be described by observations and also after analysing samples in laboratory. However, choice of properties on priority should be such which describe a sedimentary rock uniquely. Such properties yield a complete description and a rock is described in its entirety. Such properties that define the rock uniquely are referred to as fundamental. However, to select a set of fundamental properties, Griffiths (1961) had suggested the following two criteria:

(*i*) the set must contain all properties necessary to define a rock specimen uniquely, and
(*ii*) the set must not contain more properties that are necessary for a unique definition

Any sedimentary rock consists of detrital grains or particles of different types. It, therefore, follows that description of these constituent particles remains a primary task. On the basis of such an approach and also, on the criteria suggested by Griffiths (*op.cit.*), Blatt, *et al.* (1980) have come out with the following five fundamental properties:

(*a*) composition, i.e. the kinds of grains and their abundance;
(*b*) size of the grains or sediment particles;
(*c*) shapes of the grains;

(*d*) orientation of the grains, and
(*e*) packing of the grains.

Describing a sedimentary rock with these five fundamental properties leads to defining texture of that rock. Conventionally, a distinction is made in sedimentology between texture and structure. Structure is considered to be derived from variations in fundamental properties and is also regarded as large scale expression in variations of these fundamental properties. Any sedimentary structure, therefore, is a derived property of the sedimentary rock. Ripple marks, cross-bedding, etc. are therefore derived properties. Apart from such structures, there are some other derived properties, which are significant and more important than any of the fundamental property. These are mainly porosity and permeability; which are of utmost importance; especially in the field of petroleum geology. Porosity of a sedimentary rock results due to variations and spatial distribution of grain size; while permeability depends on fabric of the sedimentary rock. Estimation of these properties is important to know the reserves and availability of the energy resources. Primary sedimentary structures like ripple marks, cross-bedding, etc. are also derived due to orientation of grains in space, which is a fundamental property.

2.7 MASS PROPERTIES OF SEDIMENTARY PARTICLES

When mass properties of sedimentary particles are required to be investigated, characters of sedimentary grains and the way in which they are related to each other in space are studied. Such an arrangment is referred to as rock fabric. As stated earlier, presence of coarse or medium or fine sized particles and their orientation and packing decipher porosity and permeability of a sedimentary rock. Another important aspect is orientation of sedimentary grains, which, in turn, depends on their shape, i.e. whether sedimentary grains are spherical, ellipsoidal, platy or bladed.

Most clastic rocks are in part clastic, as the detrital or clastic grains are bound together by chemically precipitated cement. Clastic particles are classified on the basis of size. Particles larger than coarse silt size are termed as 'grains', while those finer than coarse silt (particles of mud) size are designated as, 'matrix'. Deposition of sedimentary grains of such sizes gives rise to voids or openings in intergranular space; which are later filled by chemically precipitated cement.

Post-depositional mechanical and chemical processes modify original textural characters of clastic grains and in turn, therefore, of clastic rocks. This is in the realm of diagenesis. Such a modification may take place either by fracturing or by plastic deformation of grains or due to overgrowth. Original mass properties of clastic grains may even change with time, as recrystallisation may result due to elevated temperature and pressure conditions, which is a post-depositional diagenetic effect.

3 CHAPTER

The Sedimentary Cycle

3.1 INTRODUCTION

As stated earlier in Chapter I, sediment is a regolith and that has undergone transportation. It is a product of erosion and weathering of pre-existing rocks; which in themselves are either igneous, metamorphic or even sedimentary rocks. Metamorphic rocks are the derivatives of earlier, either igneous or sedimentary rocks. Closer examination of the continental rocks shows that they are not primitive, but have been derived from sediments, produced from yet earlier pre-existing rocks. Geological history of evolution of continents is, therefore, very complex and long and indicates the continual operations of processes, under what has been termed as "Geologic Cycle"

A basic concern in sedimentology, therefore, is the origin, volume and rate of supply of sediments, produced by weathering of pre-existing rocks, and transported finally to the basin of deposition. This gives rise to the fundamental concept in understanding the operative mechanism behind the processes of uplift, erosion and sedimentation. This concept has been understood earlier by James Hutton, who has stated in his 'Theory of the Earth', published in 1788, that material is continually being shifted by erosion and transport processes from parts of the crust that undergoes uplift to parts that undergoes subsidence. In this part, sediment is accumulated and buried and even in some cases to great depths in the crust. The sediment is then changed to sedimentary rock by diagenesis or to metamorphic rock by the processes of metamorphism or even they may undergo melting to get emplaced later as igneous rock, all of which in turn are uplifted again to supply material to some other areas of deposition.

There is a continuum in these processes, which operate simultaneously and constitute what is termed as the 'Geologic cycle' or the 'Sedimentary Cycle'.

Weathering of pre-existing, source rocks, and origin and further transportation of sediments are carried out mainly by the agency of water. Though, there are other agents of weathering, significant role of water, however, cannot be ruled out. Movement of water through gullies, streams and rivers is in itself a part of 'Hydrologic cycle', which, on the other hand, also controls the 'Geologic cycle' or the 'Sedimentary cycle'.

3.2 HYDROLOGIC CYCLE

The term 'Hydrologic Cycle' refers to movement or circulation of water through atmosphere, lithosphere, hydrosphere and back to atmosphere. The concept of hydrologic cycle is introduced earlier in school. It is presented diagrammatically in Fig. 3.1.

The earth's surface receives precipitation from atmosphere, in the form of rain, snow or hail. As it falls on the ground, it finds different ways before finally meeting to ocean, from where through evaporation, it enters again into atmosphere. Depending upon the atmospheric conditions, i.e. temperature-pressure; it is circulated in atmosphere and then falls on the ground by way of precipitation.

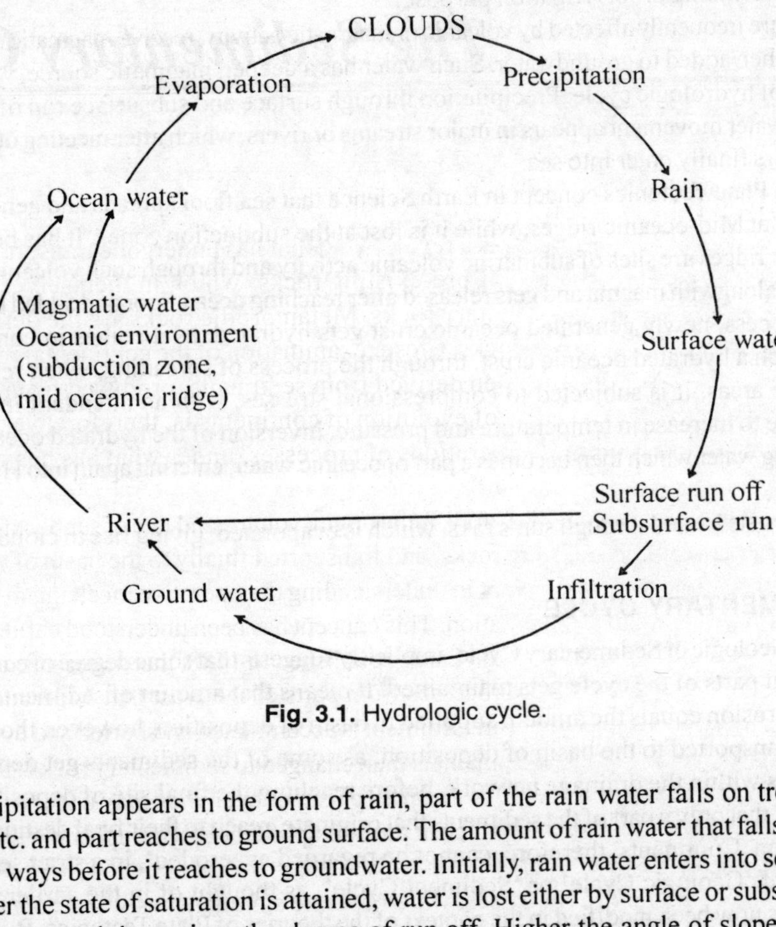

Fig. 3.1. Hydrologic cycle.

As precipitation appears in the form of rain, part of the rain water falls on trees, buildings, premises, etc. and part reaches to ground surface. The amount of rain water that falls on the ground finds many ways before it reaches to groundwater. Initially, rain water enters into soil and sub soil layers. After the state of saturation is attained, water is lost either by surface or subsurface run off. Slope of the ground determines the degree of run off. Higher the angle of slope, greater is the amount of run off and less is, therefore, the recharge/percolation to ground water.

Apart from surface or subsurface run off, rain water is also lost due to evaporation from the ground surface or soil. Rate or extent of evaporation depends on the amount of heat received from sun. Altitude or elevation of the place with respect to sea level is another controlling factor for the rate and extent of evaporation.

Trees, vegetation or forest cover on the ground is another deciding factor for water losses. Roots of the plants absorb water from soil and subsoil. Part of the absorbed water is fed to plants, and through their leaves, part is lost by transpiration to atmosphere. Apart from such losses of water, it is absorbed by soil particles due to surface tension forces and is adhered to soil particles.

If there is enough precipitation, excess over the losses due to surface and subsurface run off, evaporation and evapo-transpiration and also for soil water, it reaches to groundwater; where it occurs in the form of either perched aquifer or unconfined or confined aquifers. Groundwater movement occurs, after the hydraulic gradient is established. Once the groundwater movement starts, it generally follows the topography and comes out either in the form of springs or meets to streams or rivers. Sub-surface run-off also follows topography and under favourable conditions appears as springs or meets streams or rivers.

Groundwater in deeper aquifers may get locked up at depth, unless and until such aquifers are tapped either for drinking or for irrigation purpose.

Areas which are frequently affected by volcanic or magmatic activity, receive magmatic or juvenile water, which is then added to groundwater. Such water has a deeper, magmatic source, which later becomes a part of hydrologic cycle. Precipitation through surface and subsurface run off and also through groundwater movement appears in major streams or rivers, which after meeting other major streams and rivers finally enter into sea.

It is under the Plate Tectonics concept in Earth Science that sea floor spreads and generation of sea-floor occurs at Mid-oceanic ridges, while it is lost at the subduction zones. It has been noted that Mid-oceanic ridges are sites of submarine volcanic activity and through such volcanic activity, water comes out alongwith magma and gets released after reaching ocean floor and added to oceanic water. In the process, newly generated oceanic crust gets hydrated, as it comes in contact with ocean water. Such a hydrated oceanic crust, through the process of spreading, when reaches the subduction zone areas, it is subjected to compressional stresses, with an attendant increase in temperature. Due to increase in temperature and pressure, inversion of the hydrated oceanic crust results in releasing water which then becomes a part of oceanic water, entering again into Hydrologic cycle.

The sea water gets heated through sun's rays, which is evaporated, giving rise to clouds and the cycle continues.

3.3 THE SEDIMENTARY CYCLE

The concept of 'Geologic or Sedimentary Cycle' implicitly suggests that some degree of equilibrium amongst different parts of the cycle gets maintained. It means that amount of sediments derived by processes of erosion equals the amount supplied to basin of deposition, however, though total material is not transported to the basin of deposition; as some of the sediments get deposited at favourable places within the drainage network, before reaching the final site of deposition. It is therefore evident that only a part of the sediments that originate, reach to their final destination, i.e. basin of deposition. Continents, therefore, cannot be regarded as 'eroded', in a strict sense.

The concept of 'Geologic Cycle' or 'Sediment Cycle', as thought of in the earlier years of James Hutton, has now been modified in the context of the theories of Plate Tectonics. By the term 'Geologic Cycle' and as indicated in Fig. 3.2, implies recycling of the continental rocks. It has now been observed and proved by instrumental analysis that recycling of rocks does not occur within continents but it occurs also even within ocean basins. In the cycle are involved not only the rocks of the continents or of the crust but parts of the upper mantle also are involved. As per the concept of plate tectonics, new oceanic crust is generated at a rate of about 2 cm/yr. at the divergent plate margin and at the oceanic ridges. Newly generated crust reacts with sea water and later gets covered by gradual thickening of blanket of sediments. These sediments are progressively deposited in deeper water. As the crust cools and moves away from oceanic ridges and travel towards the subduction zone, sediments get deposited in trenches, as their final site of deposition. As the oceanic crust moves down the subduction zone, it gets heated, part of it gets melted or reacts with the sarrounding mantle, finally giving rise to volcanic or plutonic rocks. Islands or island arcs are associated with such subduction zones, which comprise either plutonic, volcanic or sedimentary rocks. Such topographic features of the crust are above the mean sea level, getting exposed to processes of erosion and denudation, thereby next cycle of sedimentation begins.

As stated earlier, during denudation of the continental masses, sediments are transported through streams and deposited in drainage basins but are not entirely removed from the continents. Continents, in a strict sense, therefore, do not get eroded. It is imperative, therefore, to look for nature of sediment load, transported through streams.

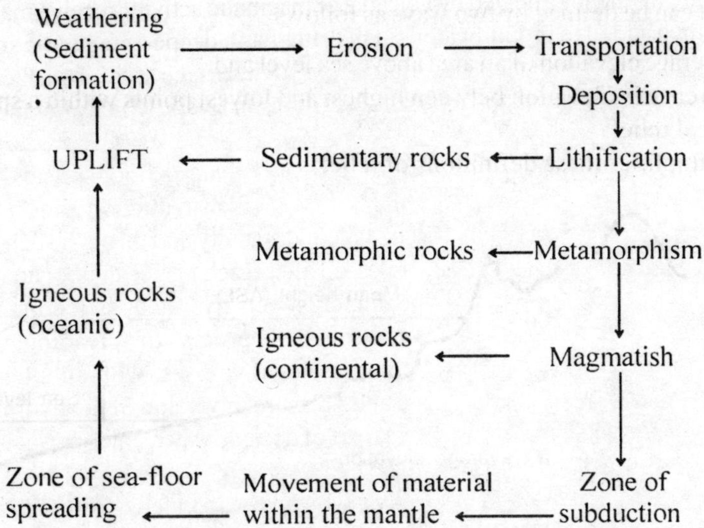

Fig. 3.2. Sedimentary cycle.

Load of Streams: Different agents of denudation act on land surface, i.e. on the continental masses. Of these different agents, marine agency is one which constantly erodes land areas bordering seas, and removes sediments to basin of deposition. Nearly 2/3rd of the earth's surface is occupied by water that is present in different seas and oceans. However, studies have indicated that proportion of sediments originated due to coastal erosion and transported to basin of deposition are negligible as compared to that derived from land surface. Many of the studies and discussions related to origin and transport of sediments, therefore, centre on the material derived from land areas of continental masses and transported to basin of deposition.

The total material, other than water, that is transported by stream is defined as 'load' of stream. Transport of such load is of three types; viz, solution load, suspension or wash load and traction or bed load. The rate at which sediment is transported through stream is called as sediment discharge, which is defined as, the sediment load moved past a given cross-section of the stream in unit time.

Considering the three different types of load, it is evident that estimation of sediment load and rate of movement are relatively easy for solution load and also for suspended load. However, it is rather difficult for bed load. Yet, estimation of bed load can be carried out with the help of sediment traps. In general, after estimating solution load and suspension load, traction or bed load is conventionally considered to be a certain fraction or commonly 10% of the solution and suspended load.

Nature of sediment load and sediment discharge vary from stream to stream and also from place to place. There are four main variables which control type of sediment load and discharge. These variables in order of decreasing importance are relief, climate, vegetation and bed rock geology. While studying denudation rates, a combined effect of all these variables is required to be considered, though the relationship amongst these variables has been poorly understood. However, some studies, carried out on estimation of denudation rates to understand the control of variables, are discussed in the following paragraphs.

Relief: Elevation difference in topography between two points or locations is generally referred to as, 'relief'. It can be defined by two ways as follows:

(*i*) an average elevation of an area above sea level and
(*ii*) difference in elevation between highest and lowest points within a specified area; termed as local relief.

Fig. 3.3 simplifies these definitions of relief.

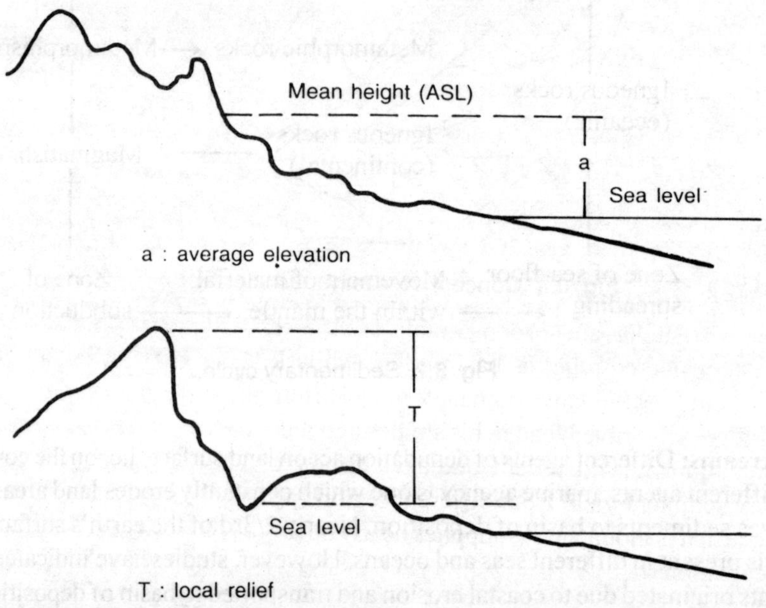

Fig. 3.3: Concept of relief (after Ruxton & McDougall, 1967).

Considering relief as a major control, studies on denudation rates have been carried out. Amongst these studies, Ruxton and McDougall (1967) have worked out the relationship between relief and sediment discharge; while considering relief as local relief. In their studies, local relief is defined as the vertical distance between major ridge crests and adjacent major valley bottoms (Fig. 3.3). An advantage of this study is that bed rock geology remaining uniform, its effect on sediment discharge can be regarded as constant. The denudation rates calculated by them indicated a linear relationship with relief. Such a relationship can be expressed by the empirical formula as:

$$QS \propto RL$$

where, QS is sediment discharge, and RL is local relief.

Another study of denudation rate has been carried out by Schumm (1968), by selecting relatively smaller drainage basins from the United States of America. However, in his studies, he had employed a different concept of relief, which he defined as the ratio of maximum elevation difference in the basin to basin length (Fig. 3.4). According to this figure, '$h_1 - h_2$' is the maximum elevation difference within the basin, while 'l' is the length of the basin. Thus, according to Schumm (*op.cit.*), relief can be expressed as, $(h_1 - h_2)/l$. His studies on sediment discharge or yield i.e. denudation rates have indicated exponential relation with relief, which can be expressed by the empirical formula as:

$$QS \propto [(h_1 - h_2)/l]^n$$

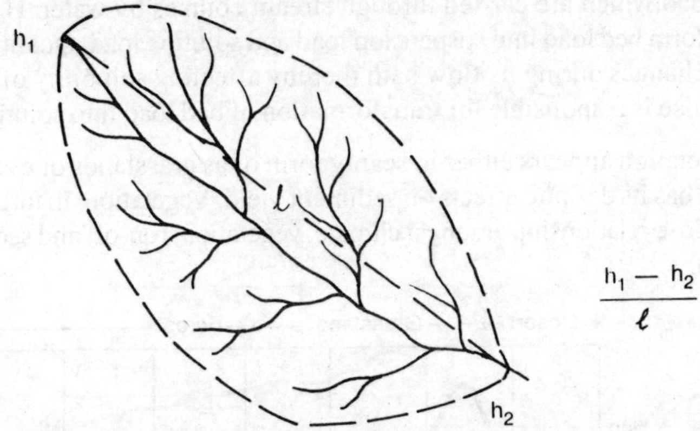

Fig. 3.4. Concept of relief (after Schumm, 1968).

However, these studies are not comparable, as different definitions of relief have been assumed. Schumm has considered relief as basin relief; while Ruxton and McDougall have considered relief as local relief. At the same, both have made different use of definitions of erosion. Schumm has not considered dissolved load in his studies and therefore, the relationship between sediment yield or denudation rate with relief worked out by both of them are not comparable.

Studies on sediment yield from different regions also indicate that nearly 5% to 10% of the earth's surface that is mountainous supplies nearly 80% or more of the clastic sediments to modern basins of deposition.

Climate: Next to relief in order of importance is climate, in controling sediment yield and denudation rate. It is evident that main meteorologic control on climate is latitudinal, which is correlated with major features of atmospheric circulation. Such an atmospheric circulation varies from equator to pole. On the basis of such circulation, latitudinal belts of relatively low and high precipitation have been identified, as follows:

(*i*) Belts of relatively high precipitation between 0° and 5° (60" to 80"/yr) at present i.e. 150 cm to 200 cm/yr. and between 50° and 60° (10" to 20"/yr) i.e. 30 cm to 50 cm/yr. at present.

(*ii*) Belts of relatively low precipitation between 5° to 30° (0" to 5"/yr) i.e. 0 cm. to 12 cm./yr at present and between 70° and 90° (5" to 10"/yr) i.e.12 cm. to 30 cm/yr at present.

Sedimentation rates are accordingly correlated with these belts of relatively low and high precipitation. Studies on particle fluxes in the western, central and eastern Arabian Sea have been carried out by Balakrishnan Nair, *et al.* (1999), between the lat. 14°13' N and 16°20' N. Their studies have indicated high fluxes during SW monsoon in the western Arabian Sea and during NE and NE-SW inter-monsoon period in the eastern Arabian Sea. It has been shown that particle fluxes are dominated mainly by carbonate during early SW monsoon but by biogenic silica during late SW monsoon. However, apart from the correlation between sedimentation rates and belts of precipitation, presence and extent of distribution of continental masses in these belts also is a deciding factor of sediment yield. At the same time, it is to be noted that present sedimentation rates cannot be applied to the geologic past, as continental masses have changed their places through geologic time, as envisaged under the Plate Tectonics concept. Therefore, present day sedimentation rates are to be employed with caution in geologic past.

As stated earlier, there are three different types of loads viz., solution load, suspension load and traction or bed load, which are carried through stream courses by water. Heat and water are the factors that transform bed load into suspension load and solution load. Quality of water, i.e. Eh-pH of stream water changes during its flow path thereby affecting solubility of minerals. Change in quality of water also is responsible for transformation of bed load into solution load.

Vegetation: Vegetation appears either in scanty form or as grasslands or even extending as thick cover as forest. It has its definite effects on sediment yield. Vegetation, in turn, depends on climate and therefore, a close relationship amongst climate, vegetation, run-off and sediment yield becomes evident (Fig. 3.5).

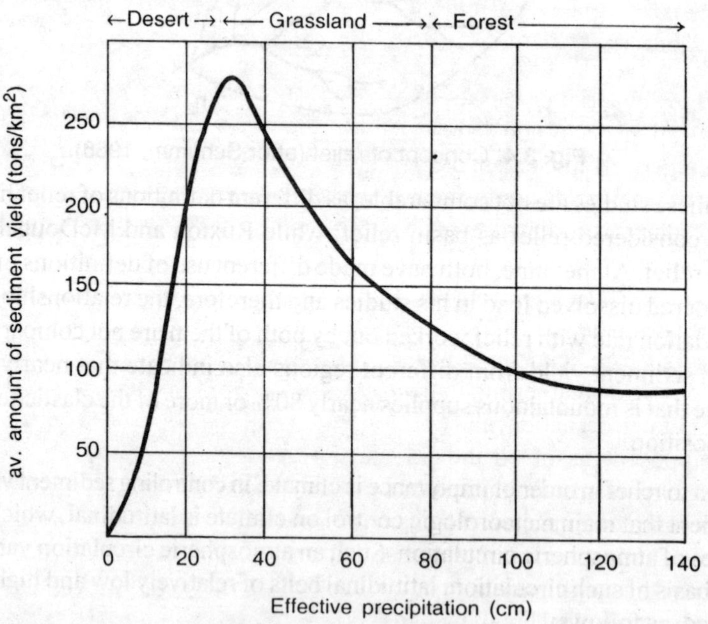

Fig. 3.5. Relationship amongst sediment yield, vegetation and effective precipitation.

Erosion and sedimentation are slow processes, which operate on the entire continental cover. These bring out changes in shape and size of land mass. Vegetation also plays an important role by producing material for erosion. Initially vegetation cover retards the rate of erosion, but does not stop erosion completely. Action of the agent of weathering, supported by gravitational forces decipher rate of erosion and therefore sediment yield.

Studies on sediment yield from the area with varying slope from 5° to 30° have been carried out, which indicated that there is a direct relation between the rate of erosion and the sine of the slope angle.

Studies on sediment yield from different river basins have also been carried out in different ways, such as; measurement of sediment discharge through a stream or amount of sediment deposited in a reservoir over a specific length of time. By noting the total amount of sediments deposited in a unit time and averaging it over the entire area of the basin, including those of the contributing streams, rate of erosion for the same area can be estimated. It can be considered as the rate of erosion under the given set of climate, and geologic and geomorphic conditions.

The amount of suspended load per unit area of a stream is controlled by a number of factors. However, it is observed that a positive correlation exists between rate of erosion and precipitation, keeping the size or area of the basin constant. Langbein and Schumm have carried out such studies at a number of sediment gauging stations. These studies have ended with a conclusion that maximum rate of erosion reaches in areas of limited rainfall of about 25 cm/year and decreases rather in very arid and also in very humid areas. Their studies can be represented graphically (Fig. 3.5), wherein sediment yield in tons/km^2 has been worked out in relation with nature of land cover and effective precipitation, which is defined as precipitation necessary to produce a given amount of run-off The studies on erosion rate have also been carried out by Judson and Sitter in 1968 from United States of America. From these studies, they have concluded that though the rate of erosion increases with decrease in discharge per unit area, it does not increase quite as rapidly as the detrital load increases. They have further observed that it is due to decrease in absolute dissolved load with decrease in discharge per unit area and an inverse relationship exists beween solid load and dissolved load.

3.4 WEATHERING

Any rock exposed to subaerial processes is subjected to weathering. It is the total effect of such processes that bring about decay and disintegration of rocks. However, in such subaerial processes of weathering, no large scale movement or transport of the weathered products of rocks is involved. Once the products of weathering are formed, they are subjected to gravity and therefore, slip from their original place and start sliding down the slope, often aided by the lubricating action of water. As the products of weathering are removed, fresh rock surfaces are agian exposed to processes of weathering. Erosion, more or less, is a synonym term to weathering and no clear cut distinction between the two can be made.

Weathering, therefore, can be regarded as a geological process. It is of two kinds; (*i*) mechanical or physical weathering. In this process, rocks are disintegrated due to temperature changes, organisms and frost action; and (*ii*) chemical weathering, in which rock constituents, viz; minerals are decomposed, dissolved and gradually get weakened by action of water, and even by atmospheric gases. Organisms and their decay products bring about chemical changes in rocks by affecting minerals. Thus, the mechanical, chemical and biological agents together bring about the weathering of rocks. Products of weathering, therefore, are the resultant effect of all these processes which act in either of the combination.

Disintegration or mechanical breaking of rocks takes place when such stresses become stronger than the strength of rocks. However, the processes of decomposition gradually reduce the strength by chemical decay of minerals and also by mineralogical bonding along the mineralogical contacts. Structural characters of the rocks, such as; joints, planes of foliation and schistosity, bedding and fractures are the easy channels for waters to move through; for plant roots to penetrate, thus facilitating further disintegration and decomposition. The extent and products of weathering, however, differ from place to place. Climatic conditions, topography—mainly slope characters and nature of the surface and types of rocks are the controlling factors on weathering. Ultimate products of weathering are broken fragments of various sizes of minerals and rocks. Soluble decomposition products are removed in solution, while residual decomposition product remains in the form of clay.

3.4.1 Mechanical Weathering

In this process, distintegration of rocks takes place, giving rise to fragments of various sizes, which occurs irrespective of climatic conditions. In arid and semi-arid climate, on exposure to sunlight, rocks get intensely heated, due to which thin outer shell expands and gradually separates out from the inner layer, which is relatively cooler. Repeated alternate heating and cooling of rocks develop fracture surfaces, that further help to facilitate chemical weathering. Such rocks, under natural conditions, therefore, give rise to shells and flakes, which further break down into smaller fragments.

Mechanical weathering readily takes place, especially when rocks are bedded and jointed. Along such surfaces separation of rocks in the form of slabs easily takes place. Water enters into pores and fractures of rocks and it freezes under cold climatic conditions. It results in an increase in volume by about 9 percent, exerting the pressure of about 2000 lbs. per square inch. Rocks are, therefore, ruptured and wedged apart, subsequently giving rise to loose and bouldery material.

3.4.2 Chemical Weathering

The process of alteration and decomposition of rocks, rather of the constituent minerals by chemical processes is termed as chemical weathering. It is brought about mainly by rain-water. Rain water as meteoric water carries with it dissolved oxygen and carbon dioxide alongwith different acids and various organic products that are generally derived from soils. Extent of chemical weathering depends on temperature, composition of rain or meteoric water, as it keeps on changing due to continual addition of dissolved mineral matter, bacteria and bacterial activity. Natural water generally dissociates into H^+ and OH^- ions, which decipher its acidity-alkalinity nature. It is generally represented by H^+ ion concentration and is referred to as pH. It commonly ranges from 0 to 14. Lower H^+ ion concentration, i.e. pH upto 7, water is of acidic character. It is alkaline water with pH greater than 7. However the value of pH between 7 and 8 denotes neutral conditions of water. It is observed that increase in acidity, i.e. lowering of pH, increases rate of weathering. It is due to dissolved CO_2, which ionises water as given below:

$$H_2O + CO_2 \rightarrow H^+ + (HCO_3)^-$$

Major chemical changes that take place during weathering are solution, oxidation and hydration or hydrolysis, and also the formation of carbonates. Different minerals have different degree of resistance to chemical weathering. Their stability is generally denoted by

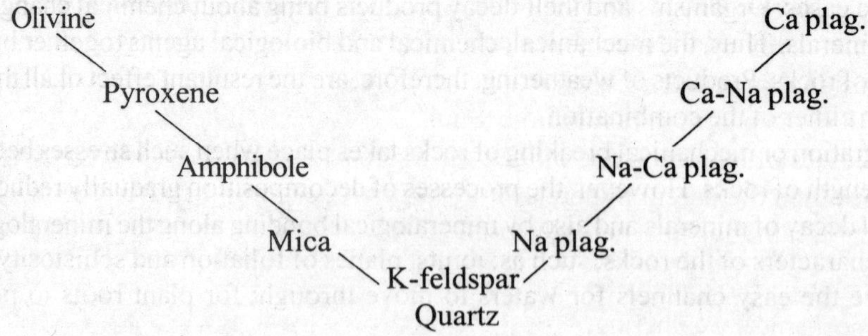

This stability order of minerals refers to the resistance of minerals to chemical weathering. Such an order indicates that mafic and ultramafic or monomineralic rocks, such as; pyroxenite, amphibolite, dunite, peridotite and the like, under given conditions of temperature and pressure undergo chemical weathering at a faster rate than the acidic rocks, such as granite, sandstone, pegmatite, having higher proportions of quartz, muscovite and potash felspar. These silicate minerals generally break down into residual products in the form of clay and soluble substances and are removed in solution. However, carbonate minerals can be removed entirely in solution.

Pure water, however, scarcely affects carbonate minerals or the rocks, like limestone and dolomite. But, due to the presence of CO_2 in water, limestone gets slowly dissolved as calcium bicarbonate—$Ca(HCO_3)_2$. Such water has low pH value, and therefore is more acidic, which brings out maximum dissolution of carbonate rocks. Impurities, like quartz and clay minerals in limestone, are left as residual products during the process of dissolution, and such products constitute the mineral matter of soils. Presence and distribution of impurities and of fractures and joints in limestone, give rise to a variety of solution features due to dissolution or chemical weathering, which are referred to as karst features.

Feldspars are common constituent minerals practically in every type of rock. During chemical weathering, feldspars are subjected to hydrolysis, under the action of slightly carbonated waters. It breaks down, as;

$$2KAlSiO_3 + 6H_2O + CO_2 \rightarrow \underset{\text{Clay mineral}}{Al_2Si_2(OH)_4} + \underset{\text{Silicic acid}}{4SiO(OH)_2} + \underset{\text{In solution}}{K_2CO_3}$$

Similarly, plagioclase feldspar also yields the weathering products. In this case, Na_2CO_3 is formed from albite and $Ca(HCO_3)_2$ is formed from anorthite. The clay that is formed is initially in colloidal solution. It consists of very small, minute particles, which are dispersed through water and which further crystallise into tiny scales or flakes. As stated, alkalies tend to pass into solution. However, soda is generally carried through solution to finally accumulate in sea water, while potash tends to lag behind and is commonly retained in the soil.

It is seen that colloidal forms of clay further break down, as follows:

$$\underset{\text{Clay}}{Al_2Si_2O_5(OH)_2} + \underset{\text{Water}}{nH_2O} \rightarrow \underset{\substack{\text{Colloidal}\\\text{Al-hydroxide}}}{Al_2O_3 \cdot nH_2O} + \underset{\substack{\text{Colloidal}\\\text{silicic acid}}}{2SiO(OH)_2}$$

Like feldspars, ferromagnesian minerals are also common constituent minerals of majority of rocks. Decomposition of such minerals can be explained, as follows:

$$Ca(Mg, Fe)(SiO_3) + CO_2 + H_2O \rightarrow \underset{\text{Silicic acid}}{2SiO(OH)_2} + \underset{\text{(Soluble)}}{\text{Bicarbonate of Ca, Mg and Fe}}$$

Quartz, i.e. SiO_2, is also another common constituent mineral, which is most resistant to chemical weathering, and therefore, is the most stable mineral.

Thus, residual products of chemical weathering remain more stable in the presence of air and water than the parent minerals from which they are formed. At the same time, as they form due to processes of hydration and oxidation, residual products contain a larger proportion of O and OH ions than the parent minerals.

Chemical weathering also helps the process of disintegration of rocks, as follows:

(*i*) initially the weakening of bonding between minerals, so that rock becomes readily susceptible to mechanical weathering;

(*ii*) by the formation of solutions, which are removed through meteoric waters, thereby imparting porosity to rock, and

(*iii*) formation of alteration products, that take place with an increase in volume from the original, that generates internal stresses, though small; due to which outer shell swells and gets detached from the fresh rock within.

Studies on weathering of khondalites, charnockites and leptynites from the parts of Eastern Ghats have been carried out by Suryanarayan and Krishna Rao (1999). Their studies have indicated that:

(*i*) climate and topography (i.e.relief) control the type and effectiveness of weathering of the parent material,

(*ii*) the climatic conditions above 600 m. altitude favour gibbsitisation/lateritisation,

(*iii*) silica is depleted more on the higher altitudes than at the foot hills or plains and

(*iv*) alumina and iron increase on the higher altitudes comparatively more than in the lower altitudes and on plain.

3.4.3 Biological Weathering

Animals and plants play an important and significant ,.role to bring out weathering of rocks and help further in the process of disintegration and decomposition.

Activity of burrowing animals like earthworms and other rodents and termites brings the mineral matter from the subsurface to surface that is removed by wind and water. These animals consume large quantities of soil to extract food and pass the indigestible particles. It has been estimated that an average soil contains nearly 1,50,000 worms per acre, which raise nearly 10 to 15 tons of fine material per annum to the surface.

Rootlets and root system of plants and trees, during their growth, pierce through the cracks, fractures and crevices present in the rock and exert an enormous pressure. These cracks and fractures are widened due to growing root system, and rocks are gradually wedged apart, thereby bringing out disintegration. Rather every type of plant is responsible for chemical weathering, as it extracts certain elements for its growth from soils and also liberate others in the process.

Water in the presence of bacteria attacks minerals and rocks more vigorously than it could do so in its absence. When organisms die, their remains decay in soil due to the activities of bacteria and fungi; thereby releasing carbon dioxide and organic acids alongwith ammonia and nitric acid in traces in soil water. The chief organic product thus generated is known as humus. Humus in soil and water containing humus acids can dissolve certain substances, such as; limonite, which are otherwise insoluble.

It has been observed that different soil biota, such as; algae, fungi, earthworms, bacteria and other types of micro-organisms acidize the soil by excreting CO_2 during respiration; while some organisms during the process of their metabolism excrete organic acids. All micro-organisms after death decompose to produce different types of organic acids. Soil scientists have observed that minerals like albite and muscovite decompose at double the rate in the presence of bacteria than in sterile clay.

Plant roots during their growth liberate large quantities of organic acids and carbon dioxide and also the absorption of inorganic substances. It is observed that CO_2 content increases downward in the soil zone, at times reaching as high as 30% of soil gases. Such a high concentration results due to absorption of CO_2 molecules on clay surfaces and within the clay structure. Studies have also indicated that rate of diffusion of CO_2 is one million times slower in water than in air.

3.5 GEOCHEMICAL CYCLE

As noted in the earlier section on weathering, any type of rock undergoes predominantly mechanical and chemical weathering. In mechanical weathering, it is simply the disintegration of source rock and products of different sizes are produced. Disintegration products are later transported and deposited at favourable sites. These various products may undergo post-depositional changes like compaction, lithification and diagenesis to give rise to sedimentary rock.

More or less, chemical weathering operates on similar lines. Decomposition of a source rock essentially proceeds with dissolution of constituent minerals and resulting products are transported through solution through fluvial system, finally reaching to ocean basins; and under favourable environmental conditions, these products give rise to chemical rocks; through the process of either saturation or oversaturation or evaporation. Resulting chemical rocks may be limestone, dolomite, Fe, Mn rich rocks or evaporites.

4

CHAPTER

Sedimentary Particles

4.1 INTRODUCTION

Previous chapter on weathering reveals that a sedimentary particle after its separation from the source rock, is subjected to varied processes of erosion and transport from its place of origin to its final destination, i.e. a place of deposition. These processes take place in different environmental set-up, like dry, arid climate, tropical-subtropical climate, glacial climate or a combination of such different climatic conditions. Transport of a sedimentary particle may also be under the influence of wind or water or ice. Topographically, it may either on land or river or fluvial system or glacier. Any sedimentary particle, therefore, is not necessarily confined to any of the climatic conditions and also to any of the transporting system. However, whenever a particle stays under a particular environmental and climatological set-up for a considerable length of time, it carries imprints of such set-up by way of physical parameters. Study of such physical characters of a sedimentary particle, therefore, has a significance, as it throws light on palaeoenvironment in which a sedimentary particle has originated.

4.2 PHYSICAL PROPERTIES OF SEDIMENTARY PARTICLES

Sedimentary particle during transport is subjected to complex processes, which substantially modify its original physical characters, such as; size and shape. Processes of attrition, wear and tear combine to alter the particle's surface. Original physical characteristics of the sedimentary particle therefore, are entirely modified. Such particle with modified characters and surface is deposited in an environment, different from which it has originated. Modifications in physical characters however, depend on mineralogical and chemical composition of sedimentary particle.

4.2.1 Sedimentary Textures

Different physical characters collectively define 'texture' of a sedimentary particle. These characteristics have a significance in sedimentological studies as they throw light on mode of transport, palaeoenvironment, characteristics of provenance or the source rocks and also palaeoclimatological conditions. Such an information when put together helps to construct depositional sedimentary environment. Physical characters of a sedimentary particle that are included in sedimentary textures are mainly; size, shape and fabric. Details of each are described in the following paragraphs.

4.2.2 Size

Grain size is a fundamental property of a sedimentary particle. Basic concept of size considers it either as a linear dimension or volume of the particle. Another consideration leads to the definition of nominal diameter, as the diameter of the sphere having the same volume as that of the particle. This may appear as an ideal concept of size. Allen (1968), however, has given three useful definitions of grain size. Sieve diameter is the most commonly used parameter, while other measures may be useful for non-quartz or quartz grains of varying shapes and density. These three different definitions are:

Sieve diameter: It is defined as a width of minimum square sieve aperture through which a sedimentary particle passes.

It may be noted, however, that a sedimentary grain passing through the overlying sieve and settling on an underlying finer sieve has a sieve diameter intermediate between two sieve values.

Volume diameter is the same as that defined as a nominal diameter.

Free fall diameter: It is that diameter of a sphere, having the same density and fall velocity of a grain in a given fluid.

While defining the concept of a linear dimension, as a size of the sedimentary particle, it is essential to define more than one linear dimension, as most sedimentary particles are irregular in shape. Generally, three dimensions are measured as: longest (d_L), intermediate (d_I), and shortest (d_S). As defined operationally, three diameters or dimensions can accordingly be measured by the sequence of steps as suggested by Griffith (1967). According to him:

(*i*) to decipher the plane of maximum projection area;
(*ii*) longest intercept across the grain and normal to this plane is the shortest diameter.
(*iii*) after deciphering the maximum projection area, to draw the tangent rectangle for the grain projection, as shown in Fig. 4.1,

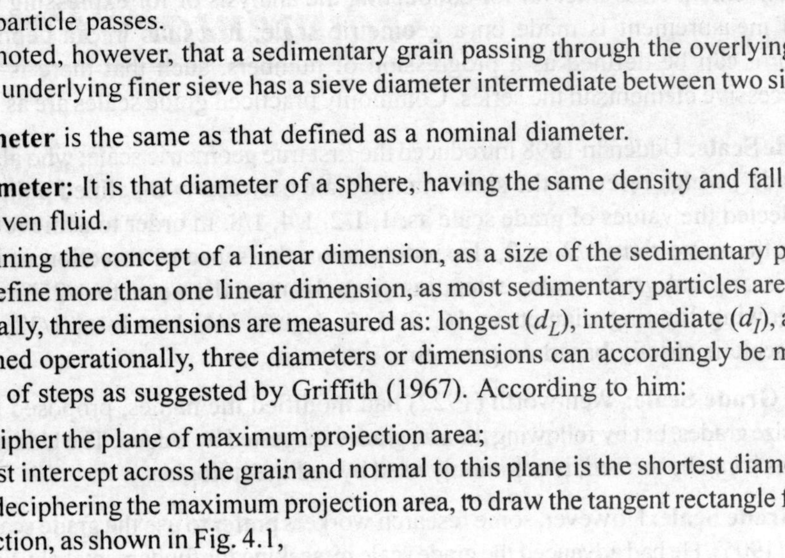

Fig. 4.1. The tangent rectangle for the grain projection.

(*iv*) long side of the rectangle is the longest diameter (d_L) and

(*v*) short side of the rectangle is the intermediate diameter (d_I) of the particle.

After defining the concept of 'size', it is to be measured on a certain scale. Each size class is accordingly termed or named. Various scales of measurements and terminology for each size class have been suggested These scales are termed as 'Grade Scales'.

Grade Scale: 'Size' of a sedimentary particle is commonly measured either on geometric or logarithmic scale, which is termed as a 'Grade Scale'. A grade scale, therefore, may be defined, as; 'an arbitrary division of continuous scale of sizes, such that each scale unit or grade may serve as a convenient class interval for conducting the analysis or for expressing the results of analysis'. If measurement is made on a geometric scale, it results into a geometric series, which, in turn, can be defined as a progression of numbers, such that there is a fixed ratio between successive elements in the series. Commonly practiced grade scales are as follows:

Udden Grade Scale: Udden in 1898 introduced the first true geometric scale; who also introduced the term 'grade', which refers to the sizes intermediate between two defined points on the size scale. He selected the values of grade scale as; 1, 1/2, 1/4, 1/8, in order to achieve a fixed ratio. By applying the same ratio 1/2 or 2, depending upon the sense of direction of the scale, he developed his original grade scale, extending from 16 mm. diameter to 1/256 mm. diameter with the following limiting diameters: 16, 8, 4, 2, 1, 1/2, 1/4, 1/8, 1/16, 1/32, 1/64, 1/128, 1/256. This scale is strictly based on geometric intervals.

Wentworth Grade Scale: Wentworth (1922) had modified the names, proposed by Udden to describe the size grades, but by following the size grades suggested by Udden. The Udden-Wentworth grade scale, as it can be termed, is commonly used by sedimentologists.

Atterberg Grade Scale: However, some research workers prefer to use the grade scale, suggested by Atterberg (1905). He had advanced the grade scale by seeking for fundamental physical properties as a basis for erecting a grade scale. His class intervals were based on the unit value of 2 mm. with a fixed ratio of 10, for each successive grade; i.e. 200, 20, 2, 0.2, 0.02. Each of these major grades was sub-divided into sub grades, chosen at the geometric mean of the grade limits. Thus, the geometric mean between 20 and 2 mm. diameter was found as, 20 × 2 = 40 = 40 = 6.32; which is rounded off to 6 for convenience. Due to this, geometric simplicity of the sub-grades was destroyed, yet keeping geometric nature of main classes.

Phi and Zeta Scales: Krumbein, in 1934 applied logarithmic transformation equation to the Wentworth grade scale and obtained the phi scale, which had class limits with integers, that increase with decreasing grain size. This scale is useful as it permits the direct application of conventional statistical practice to sedimentological data.

Krumbein had also applied -ve logarithmic transformation to Atterberg grade scale and obtained 'Zeta Scale' with properties similar to that of 'phi scale'. These transformed scales are based on the fact that any true geometric scale may be converted to an equivalent scale with arithmetic intervals, if logarithms of the scale units are substituted for diameter in mm.

Krumbein had chosen the following transformation equations:

(*i*) $\phi = -\log_2 \varepsilon$: where (epsilon) is a diameter in mm
(for Wentworth scale)

(*ii*) $\zeta = 0.301 - \log d$
(Zeta) (for Atterberg scale)

However, as larger number of points are provided on a cummulative curve than those which are on Udden-Wentworth grades, the grades are generally subdivided. Subdivision of each whole phi value into its 1/2 or 1/4 may be achieved by multiplying size values (in mm.) of the Udden-Wentworth grades by 2 and 4, respectively.

Methods of Size Analysis: Measurement of size of a sedimentary particle is carried out, usually by indirect means or methods. For the purpose, distinction between coarse and fine sediments is made at 1/16 mm (0.0625 mm), which is the lower limit of sand size on Wentworth's size classification. Coarser particles are further subdivided into smaller group that include sand and gravel, which are analysed by sieving, and larger group that includes pebbles and cobbles, large enough to be handled as individual particles for direct measurement. The distinction between two groups is chosen for convenience at 16 mm. diameter.

Analysis by Sieving: For the purpose, 6" or 8" diameter sieves, having woven wire meshes in brass rings, in US, ASTM standard or Tyler or BSS are chosen, having screen openings closest to 2 or 4 Udden-Wentworth scale. Sieves so selected are stacked on a Ro - tap sieve shaker and sieve shaking is accomplished either by hand or by electrical method for about 15 to 20 minutes. According to Folk (1955) and Rogers (1955), if sieving is carried out by following such standard techniques, it is capable of yielding results with high precision. This is considered as dry sieving. The scales which are commonly practised in sedimentological analysis and the corresponding sieve numbers are presented in Table 4.1.

Sometimes, wet sieving of coarse particles is carried out. However, generally it is not as complete as dry sieving, because film of water if formed in wet sieve prevents some particles to pass through. However, if the sediments are dirty or are partially aggregated, wet sieving often helps in separating the aggregates and obtaining cleaner products. In all cases, however, the wet sieve separates generally should be resieved through the same sieves, when dry, to remove the finer particles. Procedure that is generally followed is that the loaded sieve may be agitated in a pan of water, taking care that none of the material is washed over the edge of the sieve; or a spray of water may be directed into the sieve or a nest of the sieves.

For separation of sand size particles, sedimentation tubes have also been used, which provide rather rapid analysis. These are based on the principle that settling velocity has a greater dynamic significance than size determined by sieving. In sedimentation tubes, sample is introduced at one end of the tube and settles at the other end and the time required for different sediment size fractions to settle is measured:

(*i*) by observing the sediment as it accumulates in a narrowed end portion of the settling tube;
(*ii*) by recording automatically the pressure difference between the column of water and sediment and equivalent column of pure water and
(*iii*) by automatic recording, say by electrobalance, the weight of sediment accumulating on a pan suspended at the bottom of the tube.

Last two methods are considered to yield the results with considerable precision.

However, there are certain limitations of settling tubes for size analysis of sands, which are as follows:

(*i*) sand sizes do not follow Stokes law; but instead they fall into a range of settling velocities, which are difficult to be precisely expressed by any analytical expression;

(*ii*) it is difficult to record the tails of the size distribution with the help of any commonly used sedimentation tube. It is generally considered that information relating to genetic interpretation can be revealed only by tails of size distribution, and

(*iii*) a major limitation of the method is the number of effects produced at the time of introduction of the sample at the top of the tube.

TABLE 4.1: Wentworth and phi grade scales, along with corresponding sieve numbers

Wentworth Scale in mm.	*ϕ Scale*	*A.S.T.M.*[1] *Sieve No.*	*B.S.S.*[2] *Sieve No.*
4.00	−2.00		
3.36	−1.75		
2.83	−1.50		
2.38	−1.25		
2.00	−1.00	10	8
1.68	0.75		10
1.41	−0.50	14	12
1.19	−0.25		14
1.00	0.00	18	16
0.841	0.25		18
0.707	0.50	25	22
0.595	0.75		25
0.500	1.00	35	30
0.420	1.25		36
0.354	1.50	45	44
0.297	1.75		52
0.250	2.00	60	60
0.210	2.25		72
0.177	2.50	80	85
0.149	2.75		100
0.125	3.00	120	120
0.105	3.25		150
0.088	3.50	170	170
0.074	3.75		200
0.063	4.00	230	240
0.053	4.25		300
0.044	4.50		350
0.037	4.75		
0.031	5.00		

1. A.S.T.M.—American Soil Testing Materials
2. B.S.S.—British Soil Standards.

However, it has been observed that in spite of these limitations, settling tubes can give results of reasonably high precision and are adequate for general determination of mean size and sorting.

Thin Section Analysis: This method is generally followed, where there is a problem of disaggregation of the sample into individual grains. In such cases, thin section is prepared and size analysis is carried out, with the help of petrological microscope.

In practice, however, shapes of the particles are irregular and unknown, and the problem is to infer the original size distribution from the observed thin section. This problem is further complicated by the fact that practically all the sediments are composed of non-spherical particles, exhibiting anisotropic fabric Therefore, results of thin section size analysis depend on the direction in which the section is cut relative to the grain fabric and also on the degree of anisotropy of the fabric.

However, there is no unique solution to infer original three dimensional size distribution from the observed two dimensional size distribution. It is therefore, necessary to follow one of the two procedures:

(*i*) to adopt a standardised method for measurement of size in thin section and to consider the same as the valid method of measurement, and

(*ii*) to correct thin section size to sieve—equivalent by using empirical correction factors, that have been developed by measuring a set of 'average' sands by both methods.

In either case, it is necessary to adopt standardised procedures for thin section analysis; whereby thin section is cut in a standard orientation and grains are selected for measurement according to a well-defined sampling plan.Therefore, thin section cut parallel to the bedding; and also the method that is followed to select the grains for measurement is important, as different methods result in different weightage to size distribution. These grains are selected which are encountered on a traverse across the slide. If grains are selected at grid points, like in point-counting procedure, the grain frequencies are weighted by the volume of the grains. This method seems to be the most appropriate method of analysis, as it correlates closely with the sieve or sedimentation method, where the frequencies obtained are weight frequencies.

Separation of sediment size fractions is generally carried out according to Wentworth scale and/or also by following the phi scale in size analysis studies. Phi scale being mathematical, its suitability in calculating various size parameters, including moments, has been established.

4.2.3 Shape

Shape of a sedimentary particle is a complex property and is therefore difficult to describe precisely. Shape describes the geometric form of particles, which reflects origin, history and internal lattice structure of the particles. There are four commonly accepted main aspects of shape, which are as follows:

(*i*) Surface texture includes those aspects of shape, which appear on such a small scale that they do not generally affect the overall shape of the particle;

(*ii*) Roundness refers to sharpness of corners and edges of the particle;

(*iii*) Sphericity measures the degree to which the particle approaches a spherical shape; and

(*iv*) Form classifies particles into a number of form or shape classes that are based on certain measurements.

Another shape aspect that is gradually getting established into sedimentary textural studies is 'Fractal'. It is based on certain geometrical measurements of a sedimentary particle to characterise the particle morphology.

Various shape aspects, stated above, have been described in the following paragraphs:

Surface Texture: Transport of sedimentary particles may be either by ice, wind or water. During the transport, change in the overall shape of the particle may take place, which depends on the extent of wear and tear of the particle; yet surface of the particle bears certain marks and/or imprints on its surface in the form of various surface textures. Compared to the overall shape of the sedimentary particle, these are so small and insignificant that they cannot easily be studied under petrological microscope. Evidence of abrasion of the surface, therefore, can be examined only with the help of a scanning electron microscope.

It is seen that an interaction of the following three main factors are responsible to generate various kinds of surface textures: .

(*i*) the way the grain breaks, which in turn is a function of internal structure of the grain;
(*ii*) attrition of the surface, produced due to impact with other grains and
(*iii*) modification of the surface due to chemical action, such as dissolution and/or redeposition.

Surface textural studies are carried out mainly on quartz grains; being the major constituent, dominant mineral of the sediments, and more so, it is the most resistant to mechanical and chemical wear and tear.

SEM examination of quartz grains reveals that pits and fractures generate due to abrasion. The impact of one particle upon another gives rise to 'V' shaped depressions. It is also observed that in case of beach sands, number of 'V' shaped depressions per unit area, or density, appears to increase as wave turbulence increases from low to moderate to high. The highest densities of 'V' shaped depressions have been observed in particles of deep sea sands, which were deposited by turbidity currents; whereas grains broken by glacial ice exhibit characterisic irregular or conchoidal fracture surfaces. The grains transported by wind or aeolian sand particles show that the surfaces consist of sharp, somewhat regular, upturned plates, parallel to each other. Such a mechanical abrasion of aeolian sand is termed as 'frosting' or 'frosted finish'. Such surfaces are in contrast with the shiny, polished surfaces of sand grains from either beach or river environment.

Origin of frosting was earlier ascribed to the effects of desert dew. This dew, heavily charged with carbon dioxide, accumulates in the night on the particle surface and may serve as an agent of solution. Another view about the origin of frosting is the peripheral replacement of quartz by calcite. Such a frosting is a form of etching. Frosting can also result from sand blasting in desert conditions.

The effect of solution on the surfaces of quartz grains is also visible under SEM, in the form of solution pits, solution crevices and a variety of features produced by chemical etching along cleavage and fracture planes. It is a characteristic feature, especially in the grains from tropical climate. Such solution features on quartz grain surfaces are suggestive of high pH levels.

Striations and scratches, percussion marks and indentations or pits are varieties of surface markings. Striations and scratches are generally confined to larger size sediments, viz; pebbles, cobbles or boulders and originate mainly due to glacial action. The siliceous cobbles and coarse-grained igneous rocks, however, virtually remain unmarked and therefore ancient well-indurated tillites show few striated stones or none. Striations are generally narrow and straight, while scratches are clearly cut below the surface on which they appear and therefore are not in any way dependent on alteration of the surface adjacent to them. Nailhead scratches are striations, which have a definite head or point of origin. From this point such scratches tend to narrow or slightly taper and come to an indefinite end. If the marked cobbles are embedded in the matrix, their striations tend to be aligned with the direction of ice flow and therefore, in general, are parallel to the long axes of the cobbles.

There are four chief striation patterns, namely; parallel, sub-parallel, scatter or random and grid. Of these, sub-parallel and grid pattern appear to dominate in glacial cobbles. It has been observed that widely spaced striations of grid pattern are more common relatively on ice-jam cobbles than on glacial cobbles, as are also curved striations.

Percussion marks are impact scars and can notably be observed on pebbles of cherts and quartzites. These are caused by blows upon the pebbles or cobble surface due to impacts from other moving particles. These marks probably indicate high velocity flows.

Pits or surface indentations are produced in part by etching and differential dissolution of the surface due to inhomogeneities in composition of the rock. Coarse-grained igneous rocks are found to be characteristically pitted, whereas fine-grained rocks have generally smooth surface.

Surface markings reflect the origin of particles. However, when particles move from one environment to another and possibly back again in the earlier environment, they undergo more than one cycle of sedimentation. Particles, therefore, may undergo various kinds of solution activity, erasing earlier or old markings and new ones take their places. Reconstruction of the sedimentation history of a particle from the surface textures, therefore, is only partially attained.

Sphericity: It is the concept that has been developed by Wadell in 1932. It measures the degree to which a particle approaches a spherical shape.

Wadell has defined sphericity as the ratio between the diameter of the sphere with the same volume as that of the particle and the diameter of the circumscribed sphere (Fig. 4.2). For particles of enough large size, volume can be measured by displacement of water or mercury. Generally, sphericity is determined by measuring three linear dimensions of the particle, as discussed earlier under the heading, 'size'. If the particle is assumed to have a shape as a regular triaxial ellipsoid, volume of the particle is then

$$(n/6)\, d_L \cdot d_I \cdot d_S$$

The volume of the circumscribing sphere would then be $(n/6) \cdot d_L^3$

Sphericity of the particle, therefore, becomes,

$$\xi = \frac{d_S^3 \cdot d_L}{d_L^2}$$

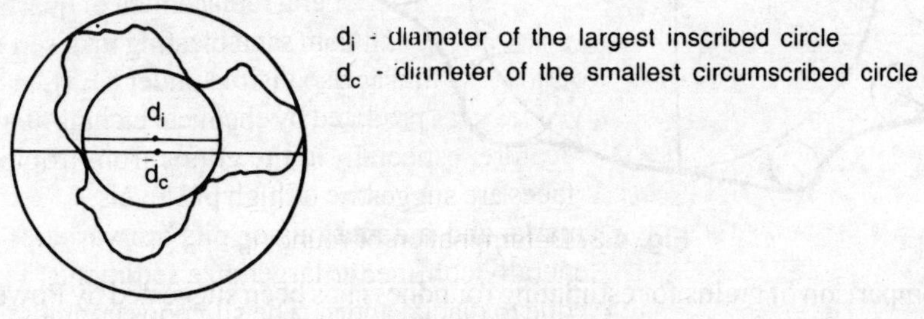

Fig. 4.2. Concept of sphericity.

However, the concept of sphericity by Wadell has been objected by Sneed and Folk (1958). According to them, though Wadell's sphericity correctly measures the degree to which a particle approaches sphere, it does not correctly express the dynamic behaviour of the particle in a fluid. They observed that particles settling in fluids or those settling on the bottom plane tend to orient themselves with their maximum area normal to the flow. Therefore, they defined the maximum projection sphericity as the ratio between the maximum projection area of a sphere having the same volume as that of the particle and the maximum projection area of the particle. The equation, therefore, comes to

$$\rho = \frac{d_S^3}{d_L \cdot d_I}$$

In most of the applications, therefore, the maximum projection sphericity is preferable to sphericity, as originally defined.

Sphericity of the particle seems to be controlled mainly by the planes of weakness, such as bedding or fractures. However, sphericity is also related to size. The larger the particle (like pebble), lower is the sphericity. In contrast, in sand-size range, sphericity increases with the increase in diameter of the sand-sized particle. However, during the transport, modifications in shape of coarse-sized (e.g. pebble) and sand-sized sedimentary particles take place. Such a size- sphericity relationship however, has not yet been well understood.

Roundness: Roundness is an attribute of a form of a particle that is related to the sharpness of edges and corners of the grain. Geometrically, it is not related to sphericity. This aspect is on a larger scale, when compared with surface texture; while it is on a smaller scale, when compared with the overall dimensions of the grain.

Definition of roundness, suggested by Wadell (1932) has been generally accepted, which states that roundness is the ratio of the average radius of curvatures of all the corners to the radius of the largest inscribed circle (Fig. 4.3). Dobkins and Folk (1970) have redefined roundness as the ratio of the radius of the sharpest corner to the radius of the largest inscribed circle.

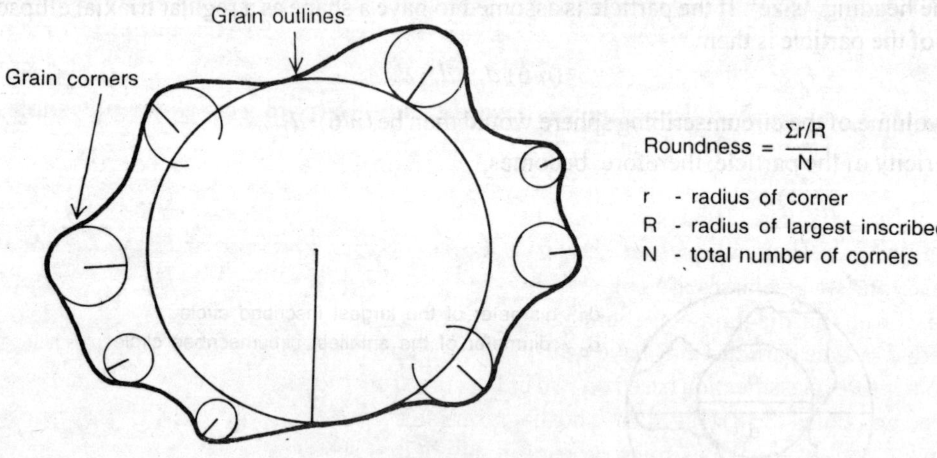

Fig. 4.3. Determination of rounding.

Visual comparison of grains for estimating roundness has been suggested by Powers (1953) and the roundness classes are given descriptive terms, from very angular to well rounded and roundness values (rho) range from 0 (zero) i.e. perfectly angular) to 6 (well rounded) (Fig. 4.4). However, accuracy of visual estimation is rather low.

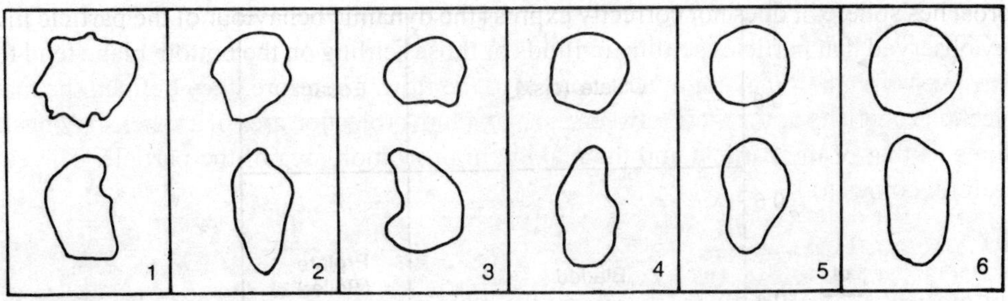

Fig. 4.4. Visual scale of rounding.

In practice, however, for precise measurements, projected image of the grain, enlarged to a suitable scale is considered. In this case, radius of each corner or curvature from the projected image of the grain is measured and radius of the maximum inscribed circle is noted. Average radius of the corners, curvatures and edges is then computed, and further divided by radius of the largest inscribed circle to obtain roundness of the grain. The equation to compute roundness, therefore, is

$$\rho = \frac{\Sigma(r_i / N)}{R}$$

where r_i = radius of every corner
N = number of corners measured, and
R = radius of the largest inscribed circle.

It has been noted that during transport, edges and corners of the sedimentary particle are gradually smoothened due to wear and tear. Sedimentary particle, therefore, becomes rounded and rounded. Thus, rounding of a particle is a function of the distance particle travels. However, no mathematical or empirical relationship has so far been established.

Form: It has been noted that shape aspects, namely; sphericity and roundness do not define shape of a particle uniquely. Several other indices have, therefore, been proposed by sedimentologists to define additional aspects. These shape indices have been based on combinations of three principal diameters of the particle.

Accordingly, Zingg (1935) proposed the use of two shape indices, viz. dI/dL and dS/dI, to define four main shape classes, as shown in Fig. 4.5, which is termed as Zingg diagram.

These four shape classes are oblate (tabular or disc-shaped), equant, bladed and prolate (rod shaped). From this diagram, it is seen that equant particles have highest sphericity, while oblate, bladed and prolate particles may have nearly the same sphericity values.

Another form classification based on two indices dS/dL and $(dL - dI)/(dL - dS)$ has been proposed by Sneed and Folk (1958) (Fig. 4.6). The diagram is further divided into a number of shape fields, by drawing lines of equal maximum projection sphericity. On this basis, different form classes that have been identified are compact (c), compact-platy (cp), compact-bladed (cb), compact-elongated (ce), platy (p), bladed (b), elongated (e), very platy (vp), very bladed (vb) and very elongated (ve).

Fractal: To define particle form, different shape parameters have so far been proposed by sedimentologists, such as, sphericity, roundness and form. These shape parameters are based on three mutually perpendicular diameters of the sedimentary particle. Various attempts have also

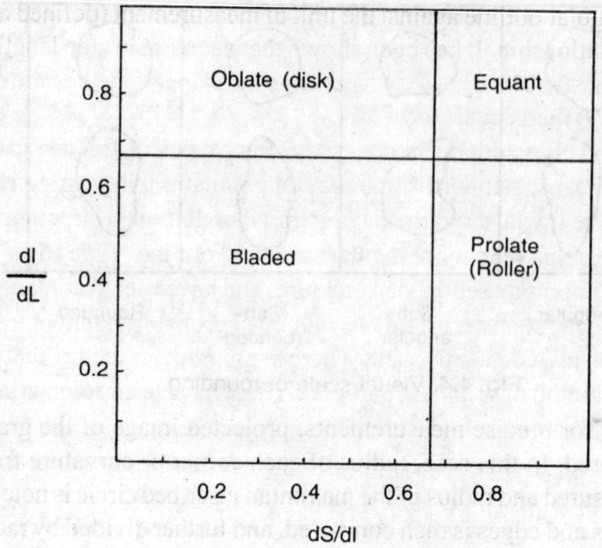

Fig. 4.5. Main classes of shape.

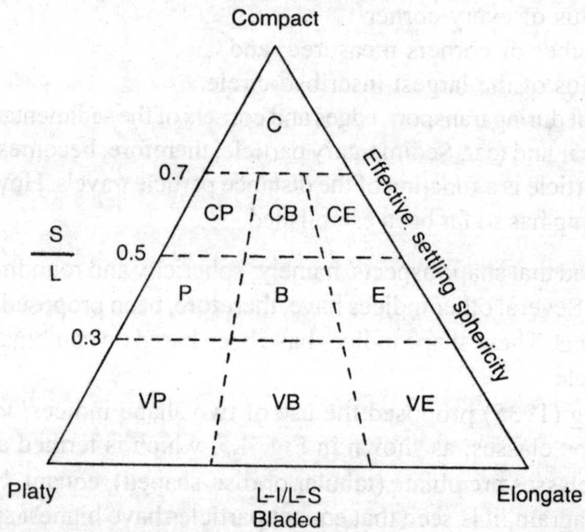

Fig. 4.6. Classification of shape aspects (after Sneed & Flok, 1958).

been made, to define shape parameters, based on the hydraulic equivalent diameter. It is, therefore, evident that considerable difficulty arises in characterisation of shape of the sedimentary particle. To overcome such a difficulty in characterisation, an additional index for shape characterisation, as 'fractal dimension' has been suggested by Mandelbrot (1977). It has been used to quantify the morphology of irregularly shaped particles by Orford and Whalley (1983). The use of fractal dimension has been made to characterise morphology of the particles with an irregular outline forming closed loops. According to the concept of fractal dimension, plot of the logarithm of perimeter estimate of an irregular outline against the unit of measurement (defined as step length) results into a stable linear relationship. It has been shown that decrease in step length results in an increase in perimeter by a constant weight (b), where 'b' is the slope of the best fitting regression line. The fractal dimension(D) then equals to $1 - b$.

Orford and Whalley (1983) have identified three different types of fractal structures on the basis of the fractal population, as, standard single fractal population, indicating regular outline morphology (Type I), converse fractal structure having irregular dominant structural nature (Type II) and irregular edge texture, relative to more regular structural outline (Type III).

Thus, these are major and important sedimentary textures and have been commonly employed in interpretation of various aspects; such as, origin of sediments, probable extent of transport, mode of transport, palaeoenvironment of deposition, etc. There are four major factors that control origin of grain size and shape distribution of sediment particles. These have been summarised by Blatt, *et al.* (1978), as follows:

(*i*) nature of the source materials, their mineralogy and texture;
(*ii*) nature and intensity of weathering, viz; physical, chemical and biological; while in bioclastic sediments, organisms including micro-organisms, bring about decay of organic substrates, boring and size reduction;
(*iii*) mechanical abrasion that involves actual breaking of grains and attrition of grain corners and surfaces and chemical corrosion of sediment particles during transport and
(*iv*) hydraulic sorting during tranport and deposition.

Studies indicate that there is strong interaction during transport between size and shape of sedimentary particles. Composition and mechanical properties, types of grains, mode of transport and deposition and mechanics of abrasion and corrosion vary with grain size and therefore have a strong effect on grain shape. The effect on shape of a grain depends on size and shape of the associated grains. It is, therefore, necessary that comparison should be made between the shapes of particles comparable in mineral composition, fabric and size grade.

4.2.4 Fabric

The term 'fabric' describes the spatial arrangement i.e. packing and also the orientation of sedimentary particles in a rock. The grain to grain relations as are observed in microsections are referred to as 'packing' in a sedimentary rock. The term packing, however, by definition, describes three dimensional arrangement of the sedimentary particles and therefore, two-dimensional observation in microsections does not describe 'fabric' adequately.

Orientation of the sedimentary particles is generally described by taking into consideration their longer axes. If the longer axes are seen to be aligned more or less in a particular direction, it is referred to as the 'preferred orientation' and fabric is described as, 'anisotropic'. Studies on grain orientation in thin section under petrological microscope, are carried out, by taking two sections; mutually perpendicular to each other. One section is cut parallel to lamination or stratification of the sedimentary rock and another section is cut perpendicular to it. Studies on fabric and grain orientation are then carried out by preparing thin section. Nature and type of the fabric is then described accordingly. Other physical properties of sedimentary rocks, e.g. porosity and permeability, magnetic susceptibility, can also be deciphered by fabric and orientation of the sedimentary particles.

In exceptional cases, constituents of sedimentary particles exhibit uniform properties or orientation in all the directions. Fabric of the sedimentary rock, in such cases, is referred to as, 'isotropic'. In practice, however, it is difficult to get such a rock. Sedimentary rocks, therefore, are characterised by the presence or even absence of fabric.

4.3 NUMERICAL ANALYSIS OF TEXTURAL DATA

As discussed in previous paragraphs, textural data related to size and shape aspects of sediments is generated by carrying out a number of measurements either on individual sediment particle or on a certain representative, fraction of sediments by weight, e.g. a sample in granulometric analysis. The data so generated is initially tabulated and graphically presented. It is then analysed, followed by interpretation. Various graphic methods are employed for presentation of the data, from which graphical parameters are computed. Interpretation related to transportation and sedimentation history of deposit is made.

Methods of data analysis are a part of statistical geology. These are described, in brief, in the following paragraphs, which are apllicable to sedimentologic data.

Sedimentologic data is either on an individual particle or a sample. Statistical inferences, drawn from such a data analysis are related initially to the sediment particle or the sample, which are then interpreted for a sedimentary deposit i.e. a litho-unit from which the sediment particle or the sample has been collected. In terms of statistics, such a lithological unit is referred to as 'population'. Thus, statistical inferences are drawn from a 'sample' to 'population'. The first step in an analysis is to summarise the data by graphic presentation and calculating a few summary statistics from the original measurements.

The main graphic devices that are employed in presentation and interpretation of sedimentologic data are:

(*i*) Histograms,
(*ii*) Frequency curve and Cummulative frequency curves, and
(*iii*) Scatter diagrams.

These are described in the following paragraphs.

(*i*) Histograms

Sedimentologic data or rather any data, is measured on a specific scale and accordingly the distribution of a parameter, within the sample is obtained. The scale on which measurements are carried out is referred to as the 'independent variable'. According to the convention in statistics, it is plotted on 'abscissa' i.e. *x*-axis; while the distribution of a parameters within the sample is termed as 'frequency' and by convention in statistics, it is plotted on an 'ordinate' i.e. *y*-axis.

Measurements on a parameter are carried out by selecting a suitable class interval of the scale of measurement, and for every class interval, frequency is counted or measured. In the presentation by way of histogram, frequency within a particular class interval is represented by erecting a column proportionate to the measurement. Frequency, therefore, is represented by way of area. Accordingly, frequencies for each class interval are proportionately represented (Fig. 4.7). There are three ways by which, frequencies are represented:

(*i*) as they have been counted or measured
(*ii*) by averaging the frequency to the total counts or
(*iii*) by converting frequency to percentages.

In histogram, trends in frequencies with respect to class interval or the position of the most frequent class, i.e. mode, are revealed more clearly than in any other graphic presentation.

(*ii*) Frequency curve and cummulative frequency curve

In the frequency curve, frequency of a particular class is represented by a point. Such points of all the frequencies are joined to construct a frequency curve (Fig. 4.8). Like histogram, frequency is represented, as it has been counted or measured or averaged or it is reduced to percentage. Thus, it is a frequency curve or a percent frequency curve. The percent frequency of one class is added to that of the immediate next succeeding class. Percent frequencies of successive classes are added upto the frequency of the last class interval. This is termed as cummulative frequency, and cummulative frequency curve is constructed.

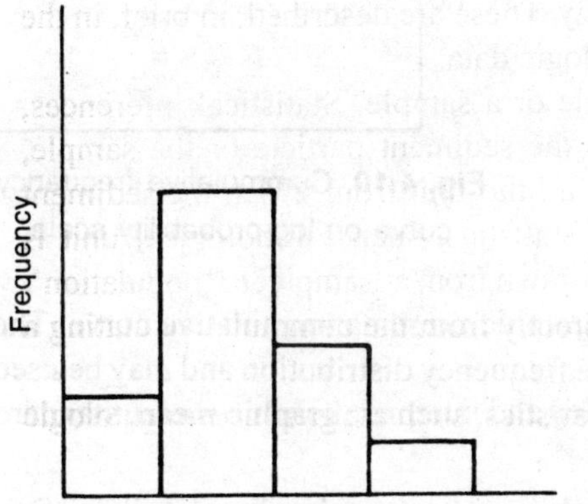

Fig. 4.7. Histogram.

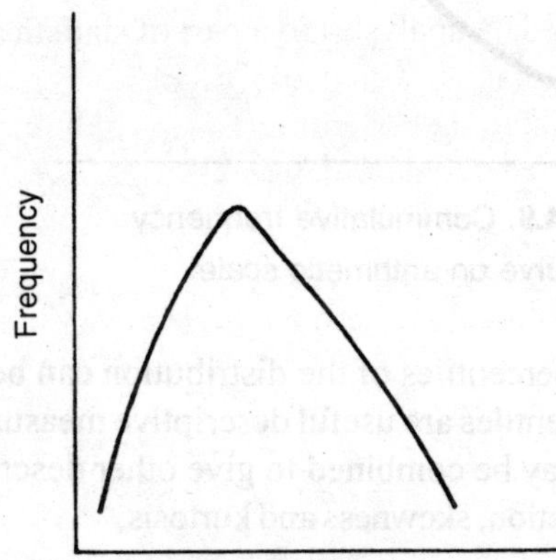

Fig. 4.8. Frequency curve.

In drawing the frequency curve, frequency of each class is represented by a point. It is represehtd at the mid-point of class interval proportional to the frequency. Accordingly, frequencies for each class are represented. These points are joined by continuous line, which generates frequency curve.

Such a frequency curve rather resembles a smoothed histogram and has certain advantages over the histogram. In a histogram, frequency is represented by an area, whereas in a frequency curve, it is represented by a point. The frequency curve is unique representation of the observations. It also permits more precise determination of mode (i.e. peak) in the frequency curve than that from a histogram.

Cummulative curves are plotted on an arithmetic scale (Fig. 4.9). The cummulative frequency of a Normal or a Gaussian Distribution, when plotted on a log-probability scale, results into a straight line. However, in practice, such a plot generally results into a three straight line segments (Fig. 4.10), as the population that is being studied is not ideally Normally distributed.

Advantages of the cummulative curves are:

(*i*) it is a unique representation of the data and may be drawn for both grouped and ungrouped data,

(*ii*) its form does not depend to any great extent on the class intervals used for grouping the data;

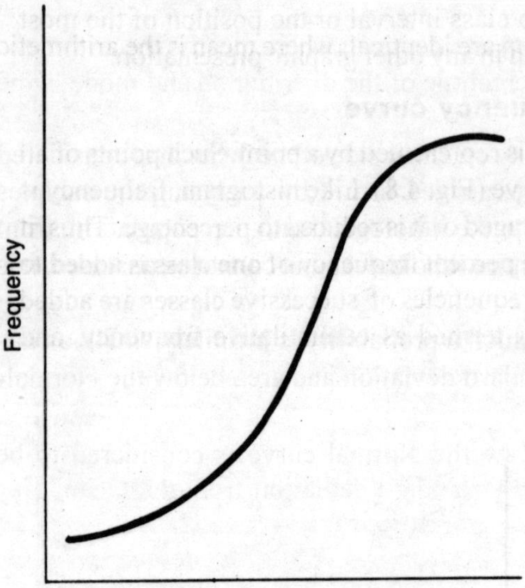

Fig. 4.9. Cummulative frequency curve on arithmetic scale.

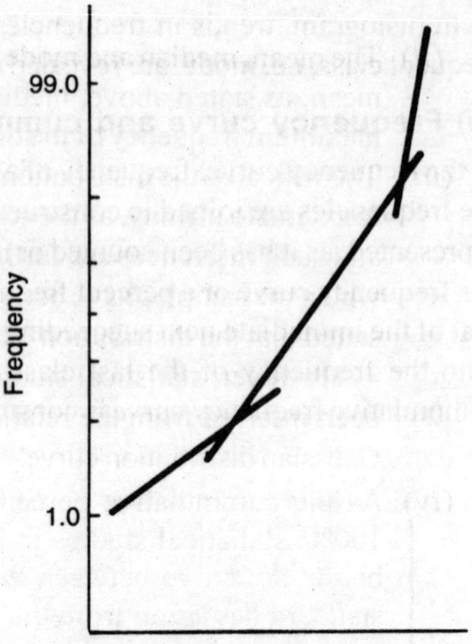

Fig. 4.10. Cummulative frequency curve on log-probability scale.

(*iii*) the percentiles of the distribution can be read directly from the cummulative curve. The percentiles are useful descriptive measures of the frequency distribution and may be used or may be combined to give other descriptive statistics, such as; graphic mean, standard deviation, skewness and kurtosis.

(*iv*) plotting the cummulative curve on probability scale permits easy visual comparison of the sample distribution with a Normal or Gaussian distribution.

Various percentiles that are read from the cummulative curve are employed to compute different graphic measures. These percentiles are so selected to have comparison with some standard and such a comparison with standard is carried out with Normal or Gaussian Distribution. It is, therefore, imperative to discuss Normal or Gaussian Distribution, which is an important statistical theory, having significance in the context of sedimentological studies.

Normal or Gaussian Distribution

Normal or Gaussian distribution is described completely by specifying its mean and standard deviation or by its square, i.e. variance. Theoritically, the normal population is infinitely very large and therefore, mean and standard deviation are to be determined from the representative samples of the population. Mean and standard deviation, therefore, are the sample statistics; which are either graphic measures or the calculated moments. The mean is a measure of the average value of the measurements, while the standard deviation or variance is a measure of the variability or the spread or sorting of the distribution in values of measurement.

Some important properties of the normal distribution are:

(*i*) The distribution is symmetrical about the mean. For a random sample from Normal distribution, the best estimate of mean is the 'arithmetic mean', which can be obtained by, $X = \Sigma\, x/n$, where '*n*' is total measurements within the sample.

(*ii*) The mean, median and mode of the distribution are identical; where mean is the arithmetic mean, as stated above, median is the 50th percentile of the distribution and mode is the maximum frequency of the distribution.

(*iii*) Theoritically, the distribution extends to infinity on either side of the mean. It is considered to extend to infinity, as the studies or inferences arrived at are by analysing the samples and such inferences are further extrapolated to the lithological unit, i.e. population, from which the samples have been drawn. As the entire lithological unit cannot be analysed, it is stated that the distribution extends to infinity.

However, statistical studies indicate that such a probability is infinitely small; as has been inferred from the relation between standard deviation and area below the Normal-Gaussian distribution curve.

(*iv*) As the cummulative percent frequency below the Normal curve is considered to be 100%, statistical studies indicate that with ±1 standard deviation from the mean, area below the curve between these limits covers 68% of the total area; whereas, with ±2 standard deviation from the mean, area below the curve is 95% of the total area; while with ±3 standard deviation from the mean, area below the curve is 99.7% of the total area (Fig. 4.11).

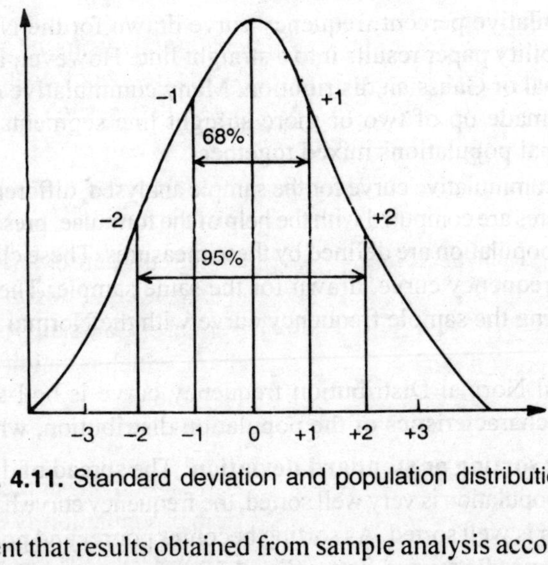

Fig. 4.11. Standard deviation and population distribution.

It is, therefore, evident that results obtained from sample analysis account nearly for 99.7% of the distribution, in the context of population studies.

In other words, it can be stated that in case of large samples, which represent Normal population, only 5% will differ from the mean value by more than 2 standard deviation. Remaining 5% of the distribution on either side of the mean are referred to as tails in the distribution, which are difficult to determine accurately, mainly because of sampling and measurement errors. In most sedimentological studies and research, therefore, 5 and 95 percentiles are used.

In sedimentological studies, number of variables are mesasured. It has been observed that such measured variables more or less approximate closely to a Normal or Lognormal distribution. However, no theoritical reason is available as to why these variables follow a particular distribution. Such variables have been listed and are presented in a tabular form.

I. Variables that tend to be normally distributed:

- Grain size in phi units,
- Grain roundness in rho units,
- Grain sphericity,
- Packing density in grains of sandstone.
- Porosity of sandstones,
- Angle of dip in cross-beds,
- Azimuths of palaeocurrent data,
- Velocity fluctuations in flow,
- Error measurement.

II. Variables that tend to be lognormally distributed:

- Grain size in mm.
- Bed thickness,
- River discharges.

(after Blatt, Middleton, Murray, 1980)

4.4. ANALYSIS OF TEXTURAL DATA

Theoretically, cummulative percent frequency curve drawn for the Normal distribution, when plotted on log-probability paper results into a straight line. However, in practice, no population ideally follows Normal or Gaussian distribution. Many cummulative curves on log probability scale, therefore, are made up of two or more straight line segments. It means that these are parts of several Normal populations mixed together.

With the help of a cummulative curve for the sample analysed, different percentiles are read and various graphic measures are computed with the help of the formulae, presented in Table 4.2. Various characteristics of the population are defined by these measures. These characteristics are reflected in the nature of the frequency curve, drawn for the same sample. These characteristics can be described by comparing the sample frequency curve with the Normal or Gaussian Distribution Curve.

Ideal or theoretical Normal Distribution frequency curve is bell-shaped. Variations in the curve depend on the characteristics of the population distribution, which are stated below,

(*i*) **Dispersion or sorting or standard deviation:** The spread of distribution about mean is sorting. If the population is very well sorted, the frequency curve has a sharp peak, indicating that population is well sorted. As sorting becomes poorer and poorer, the frequency curve gradually becomes flatter and flatter (Fig. 4.12.).

(*ii*) **Skewness:** It defines asymmetry of the distribution. In normal distribution, as mean, median and mode are identical, the frequency curve is symmetrical and skewness is zero. In case, where skewness is positive, it indicates the excess of fine particles over the coarse particles, and if skewness is negative, it indicates excess of coarse particles over fine particles (Fig. 4.13).

(*iii*) **Kurtosis:** It is the fourth measure of the population distribution and it estimates the peakedness of the distribution. However, it does not yield an additional information, that has been gathered from the earlier measures. The curve for the Normal distribution is termed

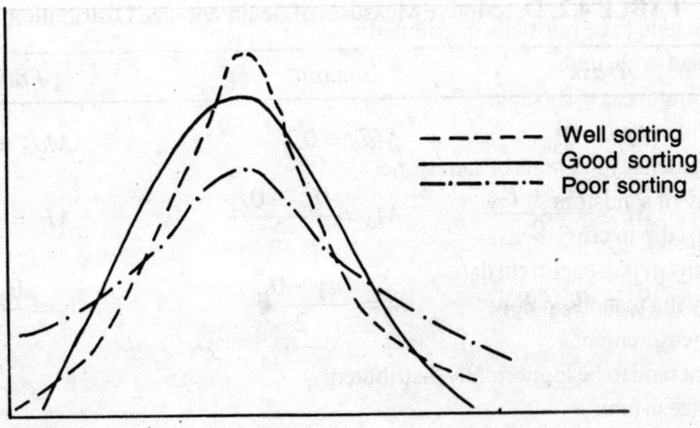

Fig. 4.12. Nature of the frequency curve and sorting characteristics.

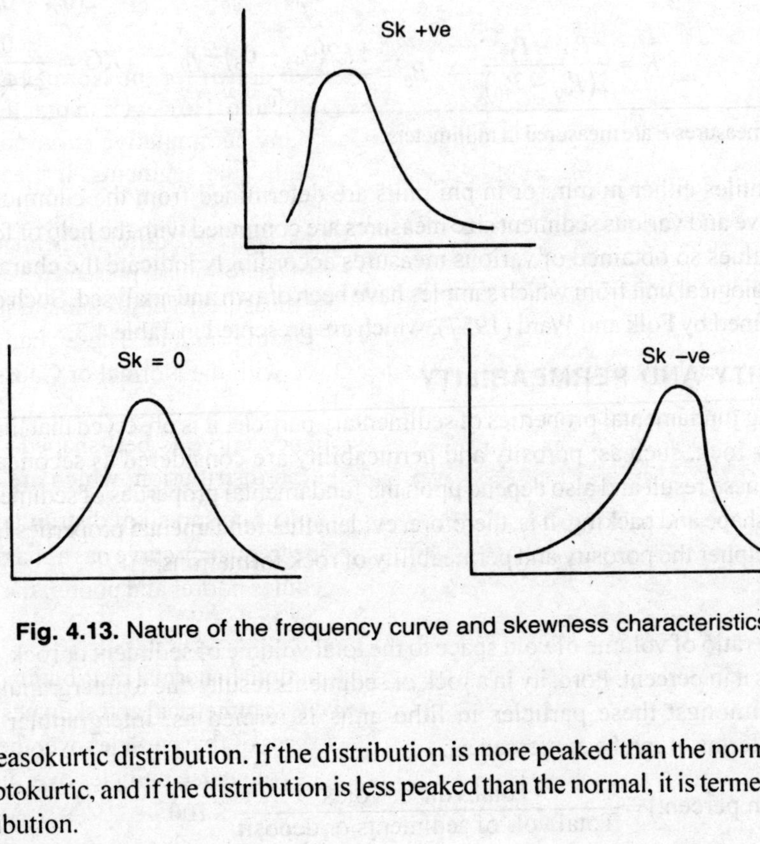

Fig. 4.13. Nature of the frequency curve and skewness characteristics.

as measokurtic distribution. If the distribution is more peaked than the normal, it is termed as leptokurtic, and if the distribution is less peaked than the normal, it is termed as platykurtic distribution.

With the help of the cummulative percent frequency curve, various graphic measures are computed with the help of the formulae, sugggestd by Trask, Inman, and Folk and Ward; which have been presented in a tabular form in Table 4.2.

TABLE 4.2: Descriptive Measures of Sediment-size Distribution

Measures	*Trask*	*Inman*	*Folk and Ward*
Median	$Md = P_{50}$	$Md_0 = 0_{50}$	$Md_0 = 0_{50}$
Mean	$M = \frac{P_{25} + P_{75}}{2}$	$M_0 = \frac{0_{16} + 0_{84}}{2}$	$M_z = \frac{0_{16} + 0_{50} + 0_{84}}{3}$
Dispersion (sorting) i.e. Standard deviation	$S_0 = P_{75} / P_{25}$	$r_0 = \frac{0_{84} - 0_{16}}{2}$	$r_I = \frac{0_{84} - 0_{16}}{4} + \frac{0_{94} - 0_5}{6.6}$
Skewness	$Sk = \frac{P_{25} - P_{75}}{Md_2}$	$a_0 = \frac{M_0 - Md_0}{r_0}$	$Sk = \frac{0_{16} + 0_{84} - 0_{50}}{2(0_{84} - 0_{16})}$
		$a_{20} = \frac{1/2(0_5 + 0_{95}) - Md}{r^0}$	$\frac{0_5 + 0_{95} - 20_{50}}{2(0_{95} - 0_5)}$
Kurtosis	$K = \frac{P_{75} - P_{25}}{2(P_{90} - P_{10})}$	$B_0 = \frac{1/2(0_{95} - 0_5) - r_0}{r_0}$	$KG = \frac{0_{95} - 0_5}{2.44(0_{75} - 0_{25})}$

The percentile measures *P* are measured in millimeters.

The percentiles either in mm. or in phi units are determined from the cummulative percent frequency curve and various sediment size measures are computed with the help of formulae given above. The values so obtained of various measures accordingly indicate the characteristics of a particular lithological unit from which samples have been drawn and analysed. Such characteristics have been defined by Folk and Ward (1957), which are presented in Table 4.3.

4.5 POROSITY AND PERMEABILITY

While defining fundamental properties of sedimentary particle, it is observed that the properties of a sedimentary rock, such as; porosity and permeability are considered as secondary or derived properties as these result and also depend upon the fundamental properties of sedimentary particle; such as, size, shape and packing. It is, therefore, evident that fundamental properties like size, shape and fabric decipher the porosity and permeability of rock formations.

Porosity

Porosity is the ratio of volume of void space to the total volume of sediment or rock, multiplied by 100 to express it in percent. Porosity in a rock or sediments results due to intergranular void space. Pore spaces amongst these particles in litho-units is termed as, intergranular porosity. In mathematical terms, it can be expressed as;

$$\text{Porosity (in percent)} = \frac{\text{Total vol. of voids}}{\text{Total vol. of sediments or deposit}} \times 100$$

The porosity of a sedimentary deposit composed of nonporous, solid particles, i.e. intergranular porosity, varies according to variations in size, shape and fabric of the constituent particles, and presence of cement, if any, between the particles. Packing of the grains is another significant aspect

TABLE 4.3: Sediment Characteristics from Statistical Measures

	Folk and Ward (1957)		*Friedman (1962)*
Standard deviation (in phi)	< 0.35	Very well sorted	0.35
	0.35 – 0.50	Well sorted	0.50
	0.50 – 0.71	Moderately well sorted	0.80
	0.71 – 1.00	Moderately sorted	1.40
	1.00 – 2.00	Poorly sorted	2.00
	2.00 – 4.00	Very poorly sorted	2.60
	> 4.00	Extremely poorly sorted	
Skewness	+ 1.00 – 0.3	Strongly fine skewed	
	0.30 – 0.10	Fine skewed	
	0.10 – 0.10	Nearly symmetrical	
	– 0.10 – 0.30	Coarse skewed	
	– 0.30 – 1.00	Strongly coarse skewed	
Kurtosis	< 0.67	Very platykurtic (tail better sorted than central portion)	
	0.67 – 0.90	Platykurtic	
	0.90 – 1.11	Mesokurtic	
	1.11 – 1.50	Leptokurtic (Central portion better sorted than tails)	
	1.50 – 3.00	Very leptokurtic	
	> 3.00	Extremely leptokurtic	

of fabric, and it directly controls the porosity of a sedimentary deposit. Studies on aggregates of uniform size, non- cemented spheres have indicated that porosity in such case varies from a maximum of 48% in open packing to a minimum of 26% in closed packing. Pores in such sediments and rocks are occupied by fluids, generally water and less commonly by oil or gas.

Other term, commonly employed to denote porosity is void ratio. It is the ratio of pore volume to grain volume. Another term is effective porosity, which is the ratio of interconnected void volume to total rock volume.

Though porosity.is considered generally for spherical particles, in nature, however, particles never attain size of sphere. There are departures from sphericity in smaller sizes of sediment particles. Such departures from sphericity result in an increase in porosity.

Pores in a rock are irregularly shaped cavities and distribution in pore sizes in a sample is therefore not regular or rather haphazard and random. An individual pore may be tubular like a capillary tube or it may be nodular or may be a thin, intercrystalline tabular opening. Walls of such pores may be of clean quartz or feldspar or may be coated with clay mineral particles or platy accessory minerals or lithic fragments. Thus, the boundary between pore space and mineral grain in sandstone is clearly defined. In sandstone, porosity varies between zero to 40%, which is more or less similar in the range of porosity variation in mudrocks. Variations in porosity of sandstone depend mainly on syn- and post-depositional changes. Presence of clay minerals and sorting of sand fraction during deposition are significant aspects, while those of less importance in affecting porosity are initial grain packing, sand mineralogy, mean size, and angularity of grains. Degree of compaction and formation of authigenic minerals are significant post depositional or diagenetic factors, responsible in reducing original porosity of sediments.

Permeability is another important secondary or derived property of a sedimentary deposit or a lithounit, which is a measure of the ease with which fluid flows through a rock. The flow of fluid through a rock or through a porous medium is a complex phenomenon, as it depends on nature of the pore space and pore space distribution, the kinds of fluid, and the pressure-gradient that drives the fluid. In addition, the flow of fluid is also influenced by its viscosity, which in turn depends on temperature. Such an empirical relationship can be expressed as;

$$V = \frac{Q}{A} = \frac{K \cdot (P_2 - P_1)}{uL}$$

where
V = apparent velocity (cm/s)
Q = quantity of transmitted fluid in unit time, i.e. discharge (cm^3/s)
A = cross sectional area (cm^2) at right angle to the direction of flow;
k = permeability (darcies, $cm^2 \times 10^{-8}$)
L = distance of flow along which pressure difference is measured (cm)
$P_2 - P_1$ = pressure difference (atmospheric) between two points, separated by a distance ($dynes/cm^2$)
u = viscosity of fluid in centipoise

The standard unit of permeability is 'darcy', which is defined as, 'that permeability which allows the flow of a fluid at the rate of 1 cm/sec. having viscosity of 1 centipoise to pass through a unit cross-sectional area, i.e. 1 cm^2, under a pressure gradient of 1 atm/cm; measured at right angles to the direction of flow'. Permeability is related to porosity in various complex ways. Porosity, in turn, is related to the way in which sediment particles of even size are packed together. This means that change in porosity is associated with the change in packing, even though the particles are of same size, and change in porosiy, in turn, affects permeability. Thus change in packing affects porosity and therefore also permeability. Distribution and variation of particles according to size in a litho-unit or a sedimentary deposit affect porosity and also permeability. In coarse-grained, well-sorted sediments, permeability is relatively high. Studies on relationship between porosity and permeability indicate that in some cases, permeability increases with an increase in porosity, while in other cases, permeability increases, but porosity remains constant. Even in some other cases, permeability decreases as porosity increases. It is therefore evident that porosity and permeability are not directly related quantities. However, in a direction parallel to any preferred orientation, permeability reaches its maximum within the fabric of a sedimentary deposit.

5 CHAPTER

Sediment Movement By Fluid

5.1 INTRODUCTION

Origin of the sediments as a part of the sedimentary cycle, is followed by transport from the place of their origin to the favourable place of deposition. Apart from the displacement of sediments due to gravity down the slope, they are transported either by wind and/or water. Sediment movement by these agents is controlled by the properties of the transporting agent in relation to sediment characteristics, and relative differences between these characteristics of the fluid and sediments decipher sediment movement and mode of transport and deposition. In the following paragraphs, properties of fluids have been discussed, followed by the factors involved in sediment movement.

5.2 BASIC PROPERTIES OF FLUIDS

The term 'fluid' indicates both liquids and gases. A fluid is a substance that can easily be deformed by a shear force. In other words, fluids have no shear strength.

On a body of fluids, both body forces and external forces act in different proportions. Fluids are set in motion when the strength of external force exceeds those of body forces. If the conditions are reverse, i.e. the strength of body forces exceeds that of external forces, a body of fluid remains at rest. Fluid consists of a number of particles and therefore, it is a body of particles that remains at rest. The body forces acting on a body of particles of fluids are termed as gravity or inertia.

5.3 CONCEPT OF SHEARING

As stated earlier, when external forces exceed body forces, fluids are set in motion. Forces acting on fluid bodies are generally resolved into two components; one normal to the surface and another parallel to the surface of the body of fluids. These are vectors, and are identified as:

(*i*) the component normal to the surface as pressure; and
(*ii*) the component parallel to the surface as shear stress.

The concept of shear or shearing is related to the movement of bodies of particles of fluids.

Shear or shearing is defined as deforming or potentially deforming condition. When such a condition is created, parts within a body of particles of fluids tend to slide over other parts along a series of parallel planes. Along these planes, slipping type movement is maximum. Thus, a body of particles of fluids is set in motion.

5.4 FUNDAMENTALS OF FLUID FLOW

In fluid particles, an internal condition of shearing stresses exists, irrespective of whether any external force is applied or not, and even an external applied force is great enough, not to cause any noticeable movement.

Another component of force is pressure, and gases respond to change in pressure. Due to this property, gases are termed as compressible fluids. Their density, therefore, does not remain constant. However, liquids behave differently to pressure. These are only slightly compressible and for the given pressure-temperature conditions, their density can be regarded as constant. These are, therefore, termed as non-compressible fluids.

Another important property of fluids that controls the way in which the fluid flows is 'Dynamic Viscosity'.

According to Newton's law, viscosity is defined as, 'the rate of deformation of a fluid in a direction normal to a surface is proportional to the force per unit area (i.e. shear stress) applied parallel to the surface'.

In mathematical terms, it is expressed as,

$$T = u\frac{du}{dy}$$

where T = shear stress, and
du/dy = rate of deformation
u = proportionality constant

The proportionality constant u is termed as dynamic viscosity. In c.g.s. units, it is expressed as gms/cm/sec.

Most common fluids, namely, air and water are Newtonian fluids, as they obey Newton's law of viscosity. But, it is true for pure air and water and at a given constant temperature, dynamic viscosity of pure air and that of water remains constant. However, water involved in sediment movement never remains pure. It gets mixed with fine or very fine material, such as clays in different proportions. Addition of such fine clayey material changes viscosity of water. Higher the concentrations of clays, higher is the viscosity of water. Substantial concentration of clays gives rise to mud flows. High concentration of clays not only increases the viscosity but it also changes the way in which the suspension, i.e. mixture of water and clays, responds to shear stress. In such cases, therefore, coefficient of viscosity does no longer remain constant and such fluids no longer obey Newton's law of viscosity; which then are termed as non-Newtonian Fluids.

When a body of particles of fluids is subjected to shearing stresses and if the intensity of shearing stresses overcomes the internal resistance of particles to shearing, movement within a body of particles results. A body of particles is able to resist shearing, if the particles are able to remain in contact with one another.

5.4.1 Reynolds Number and Froude Number

Resistance of a body of particles to shearing, however, depends on the following factors:

(*i*) Gravity,
(*ii*) Conditions within interstitial fluids, if present,
(*iii*) Proportional effect of electrostatic forces, resulting from uneven distribution of changes over the surfaces of the particles.

The effect of electrostatic forces varies with particle size. It is noticeable in particles which are smaller than coarse silt, i.e. smaller than 0.03 mm. or 5 ϕ. In such cases, these particles behave as small, tiny magnets, where the electrostatic forces are strong enough to hold the particles together, thereby increasing their shearing strength. Such bodies of fine-grained particles are termed as cohesive soils.

In case of coarse-grained particles, size of which is larger than fine sand, i.e. greater than 0.25 mm or 2 ϕ, they are not bound by electrostatic forces. Cohesion, therefore, does not exist in these particles. Such particles, therefore, can be easily eroded and dispersed, and are, therefore, termed as cohesionless particles.

Another important aspect is the collective response of particles to shearing or shearing stresses for physical transport. Due to collective response, behaviour of an individual particle cannot be analysed and also does not help to predict how the particles are to respond. Physical transport of particles is due to gravity shearing and where fluids are absent.

When shearing stresses or a tangential component of gravity operates on particles, an initial response is to resist to shearing stresses. It results due to creation of frictional forces within it. When shearing stresses exceeds the frictional forces within the particles, the particles are set in motion. Just when the movement within the particles results, the volume they occupy increases. Technically, it is known as dilatency, and density of particles at the point of flow is known as critical density.

When movement within cohesionless particles begins, particles get sheared. Its dilatancy, therefore, is termed as Dynamic dilatancy. This was first discovered by Bagnold and therefore, it is named after him, as, Bagnold Effect.

Flow of fluids or in other words, velocity of fluids depends on number of factors, such as; its density, its dynamic viscosity, distance or length involved in movement, etc. It is to be noted that water is 800 times denser than air. This indicates that there are number of variables that decide or affect the velocity or flow of fluids.

These variables, as some of them have been stated above, have some units. It means that these are the numbers with some dimensions. The variables that decide the type or characteristics of a flow are made dimensionless. Two such dimensionless numbers have been introduced to define characteristics of flow of fluid.

Dimensionless numbers are derived with respect to reference standard. Therefore, all the lengths are expressed as ratios of some reference length or all velocities as ratios of some reference velocity. For the movement of a body of particles, the viscous and inertial forces act on the particles. Therefore dimensionless numbers, as Reynolds Number and Froude Number have been derived to express the characteristics of a fluid flow.

Nature of flow as laminar or turbuent can be given by Reynolds Number, which is interpreted as being proportional to the ratio between inertial and viscous forces. In mathematical terms, it is expressed as,

$$\text{(Reynolds Number) Re} = \frac{u \cdot L \cdot \rho}{\mu}$$

where, u = reference velocity
L = reference length
ρ = density
μ = dynamic viscosity

Reynolds number, therefore, is interpreted as the ratio between inertial and viscous forces. It defines the characteristics of a fluid flow, either as laminar or turbulent. There is a critical value of the Reynolds Number at which the laminar flow turns into a turbulent flow. However, such a critical value depends on the choice of representative length and velocity and also on the geometry and other related properties of the flow system.

Laminar flow is defined as a fluid flow in which parts of the fluid slide over one another along surfaces that conform to the shape of the boundary of the fluid.

In case of the flow past a settling spherical particle, flow remains laminar and does not separate from the surface of a sphere upto a Reynolds Number 24. If the fluid flows in a pipe, it is laminar or streamlined upto Reynolds Number less than about 2000. It means the different layers or particles of fluid generally slide past each other and no irregular eddies are created. Therefore, there is no diffusion of particles from one layer of fluid to another. If the Reynolds Number exceeds 2000, fluid flow is turbulent, with generation of eddies and therefore, there is diffusion of fluid particles from one layer to another (Fig. 5.1).

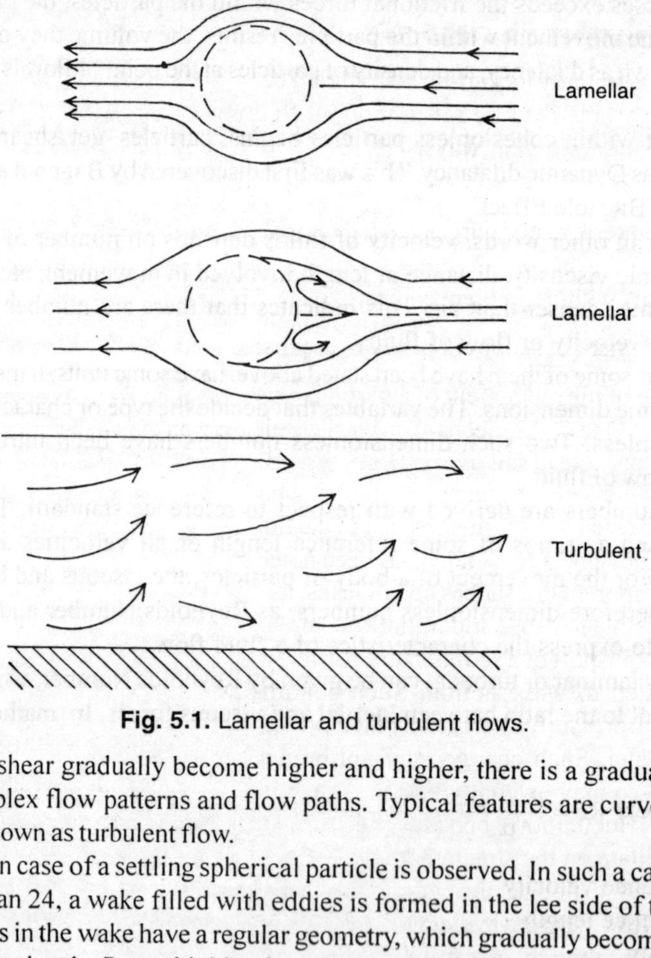

Fig. 5.1. Lamellar and turbulent flows.

As the rates of shear gradually become higher and higher, there is a gradual development of complex and complex flow patterns and flow paths. Typical features are curved and spiral flow paths, which are known as turbulent flow.

Nature of flow in case of a settling spherical particle is observed. In such a case, with Reynolds Number greater than 24, a wake filled with eddies is formed in the lee side of the particle. In the initial stages, eddies in the wake have a regular geometry, which gradually becomes more and more irregular in nature and at the Reynolds Number greater than 1000, the wake is fully turbulent.

Another dimensionless number for characterisation of fluid flow is Froude Number. It is analogous to Reynolds Number and is defined as a ratio between inertial and gravity forces. It is expressed as:

$$\text{(Froude Number) Fr} = \frac{u}{g \cdot L}$$

In case of water flowing in an open channel, having water depth '*d*'. Froude Number is given as $u/g \cdot d$, where u is the average velocity of fluid flow. In case, where water flows in an open channel, having long wavelengths compared with water depth, velocity of waves becomes equal to $g \cdot d$., and Froude Number therefore becomes unity.

These characteristics suggest that if the Froude Number is greater than unity, waves cannot travel upstream, as the downstream velocity of flow is greater than the upstream velocity of waves. This creates some fundamental differences in the type of flow. If the Froude Number is less than unity, the flow is tranquil, streaming or said to be sub-critical. However, if the Froude Number is greater than unity, nature of flow is rapid, shooting or super-critical.

These two dimensionless numbers, viz. the Reynolds Number and Froude Number thus define the flow characteristics, as; laminar, stream lined or turbulent and tranquil or rapid, respectively.

5.5 MODES OF SEDIMENT TRANSPORT

A sedimentary particle is initially locked-up on a stream bed among other particles.It is, therefore, not free to move. If a particle is to be set in motion, fluid has to energise it. This mechanism takes place in either of the following ways:

(*i*) Fluid lift force
(*ii*) Changes in fluid pressure and surges related to wave action
(*iii*) Fluid impact
(*iv*) Support from flow within turbulent eddies.

Their effects on movement of sediment particles are discussed in the following paragraphs:

(*i*) In an initial stage of the movement of a particle, two main mechanisms operate, which are due to fluid force. One of the mechanisms is the fluid drag on exposed grains, which makes the sedimentary particle to roll over other surrounding particles. Second mechanism is the hydrodynamic fluid force, because of which the particle gets lifted vertically up from the bottom and is projected into the moving part of the fluid. Such a mechanism has been observed in the sediment movement by wind. It has been noted that in an aeolian movement, sediment particles are transported in a succession of leaps, termed as saltation. It is the hydrodynamic lift force responsible to cause and initiate the movement of sedimentary particles.

(*ii*) Movement and transport of a sedimentary particle are also due to changes in fluid pressure and surges related to wave action. Such condition exists in a turbulent flow, where changes in fluid pressure often take place on short-term scales. These changes affect sediments in water. Such change may not be necessarily restricted to water/sediment interface, but even may extend within the interstitial water within the sediment particle. These pressure fluctuations beneath the turbulent current are responsible for larger particles to oscillate on the stream bed; while decrease in fluid pressure above the water sediment interface dislodges the sedimentary particles.

In a stream flow, where wavelength is more than half of the depth of water, surges or eddies are created, thereby causing movement of the sediment particle.

(*iii*) The factor of fluid impact on movement of particle however, cannot be considered separately. It is rather a part of either of the three mechanisms, responsible for sediment movement, and therefore it is a distinct factor.

(*iv*) In turbulent eddies, a factor of upward flow of water provides energy for moving particles. So long as the upward velocity is more than the settling velocity of the particles, the particles remain in flow and get transported.

Once the sediment particles are energised by fluid in either ways mentioned above, sediment particles are transported. The transport of particles takes place by two distinct modes, identified as two distinct kinds of loads:

(*i*) **Suspended load:** Current of fluid either air or water, disperses fine sediments, mainly; silt and clay with varying proportion within the main body of flow, and also in sand-sized particles;

(*ii*) **Bed load:** In such mechanism, particles move in response to several processes that operate as a result of shearing at the boundary of a flow.

(*i*) **Transport by suspension mechanism:** Sediment movement by suspension is possible because of turbulent current, which may be created in both, that is in air and water. Suspension of particles in water is termed as aqueous suspension. In an aqueous suspension, turbulence is maintained upto a point, where mixture of sediments and water has a density of about 3 gms/cm. By the addition of suspensoid, there is an increase in density, turbulent flow then can change into either a debris flow or a mud flow. In aqueous suspension, fine sand, very fine sand and coarse silt may interchange with their substrates. Limit for fine sand is provided by the settling velocity in water of 4 cm/sec. At the average forward velocity of about 50 cm/sec., fine sand remains in suspension.

In a turbulent aqueous suspension, where exchange of sedimentary particles takes place with its substrates and where the quantity and distribution of particle size results into an upward decrease in size, it is termed as a graded suspension.

In suspension, if load travels independently and with no interchange with substrate, such suspensions are independent of their substrate. There is more or less uniform size distribution, resulting into uniform suspension.

The sediment movement by suspension due to air currents is termed as aeolian suspension. Movement by air is basically governed by the same laws and principle as those for suspension by water.

(*ii*) **Transport by bed load:** Mechanism by which sedimentary particles move on the stream bed, as bed load is termed as traction.

Movement of sedimentary particles by traction has been observed, where wind blowing across dry sand shows that shearing drag activates a sheet of bed load sediment, which is pushed along at the base of air current. Such a sheet of sand is termed as 'traction carpet'. During transport by traction, particles may jump and bounce. Such particles collide and bounce at the end of their trajectories on other particles, resulting into a forward movement. This bouncing and collision movement of particles in a fluid is termed as, 'saltation'. The forward movement of particles that results due to collision of other particles is termed as 'surface creep'. Generally, 75% of the particles move by saltation, while 25% move by surface creep in an aeolian environment.

5.6 SEDIMENT SORTING AND HYDRAULIC EQUIVALENCE

When sedimentary particles are set in motion and subjected to transport by a moving fluid, different particles respond differentially to fluid movement. In transport, therefore, there results a grouping of sedimentary particles, which yield the collective response in a similar manner to fluid flow at the same time. Separation of such a group of particles takes place from other particles that behave and respond differentially to fluid flow at the same time. This means that fluid flow carries out `sediment sorting'. In such case, however, velocity of fluid flow and other physical properties of fluid are presumed to be constant

However, to the flows of different intensity, response of sedimentary particles differs. As per the variations in intensity of flow, the same particle may either slide or roll or may be transported either by saltation or intermittent suspension or even by continous suspension. It is observed that fine sized sediments having lower settling velocities than the coarse sized sediments are rather uniformly distributed. In flows with relatively high velocities, suspended sediments are more uniformly distributed than in flows with relatively low velocities.

It has been noted that in a fluvial system of moderate size, majority of grains coarser than 10 mm in size cannot be transported in the lower part, but rather accumulate as lag gravel deposits in its upper part. Sedimentary particles ranging in size from 1 mm to 10 mm generally move by rolling over the abundant sand size grains and are likely to be rapidly transported through the river system; whereas coarse to fine sand size particles move mainly by traction and intermittent suspension. Fine size particles, such as; silt and clay are transported mainly as wash load in a continuous suspension. These may get transported rapidly either to river mouth or may get deposited on a flood plain due to overbank flooding.

The term 'hydraulic equivalence' indicates that the grain population that is deposited from suspension under uniform conditions have same range of settling velocity. This concept can be tested for a sample by separating it into two or more subsamples, on the basis of mineral composition or size, shape and density. For the purpose, settling velocity of these separated particles is determined and their distribution is compared. If the settling velocity differs, it can be ascribed to either of the following alternatives:

(*i*) Original conditions of settling of particles have not been correctly reproduced in the laboratory. It is also posssible that particles initially could have been transported by wind, followed by transport in water. Settling velocities of the particles obviously differ in both the mediums.

(*ii*) It can also be possible that some of the particles have not been deposited by settling but have moved by traction or saltation.

(*iii*) Another possibility is that the sedimentary particles may get modified by post-depositional process, such as; dissolution of some particles due to later addition in a deposit through percolating waters.

Granulometric analysis of sediments yields an information about differnt size fractions present in a deposit. The plot of cummulative size frequency on an arithmetic probability graph reveals distinct breaks in size distribution. These breaks and size population accordingly have been ascribed to transport mechanisms, as, traction or bed load and saltation and suspension. This has been explained earlier in the chapter on, 'Sedimentary Texture'.

5.7 GRAIN ORIENTATIONS

Orientation of sedimentary particles during transport is measured with an assumption that the particles are ellipsoidal in shape. Orientation of the grains is measured in two dimentions, by considering their longer axes. Such a section or plane, however, includes the long and intermediate axes of the particles. Angle of dip of this plane with respect to stratification is referred to as the angle of imbrication.

In a turbulent wake, sedimentary particles orient with their maximum projection area normal to the direction of settling. In such case, Reynolds number is greater than one. Fine sized particles, however, do not form a wake and therefore fail to get oriented during settling. In this case, Reynolds number is less than one. On this basis, fissility of shales is considered as a result of orientation of particles during compaction and not as a result of depositional mechanism.

Orientation of grains in most of the sedimentary deposits is related to the mechanisms that operate during the final stages of deposition, rather than to the processes, which operate during transport. Grains that can roll on the bottom of the bed usually assume the orientation with the long axis normal to the current direction. The grains which have one end larger than the other, generally point up current.

Experiments and observations on the transport and orientation of sand grains indicate that the orientation is generally parallel to the current direction, and it is the most common orientation. Grain orientation parallel to current direction is also produced in deposits of high concentration of dispersed particles. It is only a small proportion of sand grains, which generally do not show a fairly well developed orientation in the bedding plane of a deposit.

Pebbles of disc shape generally orient with their plane of maximum projection, dipping more commonly in one direction. Such a direction is generally upstream, and orientation results into imbrication. Development of imbrication results in traction deposits, in which well sorted and platy pebbles are deposited in a matrix of smaller particles. Pebbles depositing on sand bed exhibit poor development of imbrication. River gravels exhibit a marked imbrication with upstream angle of dip varying between 15° and 30°.

Pebbles deposited on beaches orient with their long axes parallel to the shoreline; i.e. normal to the direction of swash and backwash; whereas sand size grains orient parallel to the backwash. In this case, the angle of imbrication is generally less than that developed in a river system.

It is, therefore, evident that analysis of grain orientation, in combination with other textural parameters helps to interpret palaeocurrent direction and thereby palaeoenvironment of deposition.

6
CHAPTER

Primary Sedimentary Structures

6.1 INTRODUCTION

Sedimentary structures are the characteristic features of any sedimentary rock, and are easily observable and identifiable visually or by a naked eye. Small-scale features that are confined more or less to a grain of a sedimentary rock describing its geometrical aspects are termed as sedimentary textures. In comparision, sedimentary structures are on a larger scale.

Sedimentary rock is a result of deposition of sediments, in a depositional sedimentary environment. Such an environment is marked as a geomorphic unit, bounded by geographical boundaries and is characterised by a unique set of physical, chemical and biological processes. These processes operate with a specific rate, for a particular period of time. Sedimentary structures develop due to such processes and, therefore, have characteristic features of environment of deposition, in which they are formed. Study and analysis of sedimentary structures therefore, throw light in understanding the processes that operated in a particular depositional environment.

Sedimentary rocks exhibit a variety of sedimentary structures, that are formed either due to physical or chemical or biological processes. Some structures form at the time of deposition of sedimentary rock, i.e. syn-sedimentary structures or after the deposition, i.e. post-depositional. Sedimentary structures are, therefore, classified on the basis of the mode and the time of their formation, i.e. during or after deposition of the sediments.

Sedimentary structures are, termed as primary, when they are formed at the time of sediment deposition, i.e. syngenetic or if they are formed after deposition they are termed as secondary or diagenetic. Biogenic sedimentary structures originate due to the acivity of organisms, during the process of deposition of sediments. This is the genetic classification. Sedimentary structures are also classified on the basis of morphological characteristics. However, each scheme has its own limitations. It is because that full description and complete understanding of sedimentary processes and depositional environment are not possible from any sedimentary structure, which are, therefore, classified on the basis of physical, chemical and biogenic processes that are responsible for their formation. The scheme of classification of sedimentary structures is presented in Table 6.1. Major and significant sedimentary structures are described in the following paragraghs.

TABLE 6.1: Classification of Structures of Sedimentary Rocks (after Pettijohn, 1974)

Inorganic Structures		
Mechanical("Primary")	*Chemical("Secondary")*	*Organic Structure*
A. Bedding geometry	A. Solution structures	A. Petrifications
1. Laminations	1. Stylolites	
2. Wavy bedding	2. Corrosion zones	
	3. Vugs, oolicasts,and so on	
B. Bedding internal structures	B. Accretionary structures	B. Bedding (Weedia and other stromatolite)
1. Cross-bedding	1. Nodules	
2. Ripple bedding	2. Concretions	
3. Graded bedding	3. Crystal aggregates (spherulites and rosettes)	
4. Growth bedding	4. Veinlets	
	5. Colour banding	
C. Bedding-plane markings (on sole	C. Composite structure	C. Miscellaneous
1. Scour or current marks (flute)	1. Geodes	1. Borings
	2. Septaria	2. Tracks and trails
2. Tool marks (grooves and so on)	3. Cone-in-cone	3. Casts and moulds
4. Fecal pellets and coprolites		4. Fecal pellets and coprolites
D. Bedding-plane markings (on surface)		
1. Wave and swash marks		
2. Pits and prints (rain and so on)		
3. Parting lineation		
E. Deformed bedding		
1. Load and founder structures		
2. Synsedimentary folds and breccias		
3. Sandstone dykes and sills.		

6.2 INORGANIC STRUCTURES

Bedding or stratification

It is a very common and universal primary structure of sedimentary rocks. The terms 'bedding' or 'stratification' are synonymous, while the term 'stratified rocks' is used for 'sedimentary rocks'.

Most of the primary structures are a variety of stratification, which are preserved more commonly on bedding planes of sedimentary rocks.

Stratum or bed refers to a layer of sedimentary rock that can be distinguished from a layer above and below due to variations in lithological characters. The term 'stratum' refers to all thickness and types of layers, whereas, the term 'bed' refers to layers, which are greater than one cm. in thickness. The layer less than one cm. in thickness is called 'laminae' (Photo 6.1).

Photo 6.1. Fine-grained sandstone showing laminations.

Scale of stratification thickness, suggested by Ingram (1954) has been commonly accepted and is used to describe characters of stratified rocks in the field. The scale is presented in Table 6.2.

TABLE 6.2: Scale of Stratification Thickness (after Ingram, 1954)

Very thickly bedded	Thicker than 1 m.
Thickly bedded	30 - 100 cm.
Medium bedded	10 - 30 cm.
Thinly bedded	3 - 10 cm.
Very thinly bedded	1 - 3 cm.
Thickly laminated	0.3 - 1 cm.
Thinly laminated	thinner than 0.3 cm.

The boundaries of a bed may be sharp or gradational. Sharp boundaries may result due to sudden change either in depositional condition or by erosion or by diagenetic alteration of an earlier gradational boundary. The lower boundary of a bed is called sole, which may display different edimentary structures, such as; sole marks, casts or moulds. Distinction of a bed or a layer from the one above and below has given rise to a concept of sedimentation unit. According to Otto (1938), sedimentation unit can be defined as, "that thickness of sediment which was deposited under essentially constant physical conditions".

Formation of the sedimentation unit takes place under the prevailing current of uniform velocity. Such a current has a capacity to carry sediments of certain size range and therefore sedimentation unit is characterised by sediments of specific size range. Current also has mean velocity that carries sediments with specific mean size. If a sudden or drastic change in velocity conditions of current result, it looses or gains the strength to carry sediments of specific mean size. Accordingly, fraction of sediments, the current is incapable to carry, separates out of the fluid systems and deposits on the stream bed. This give rise to a sedimentation unit. Accordingly, a stratum or bed is formed.

According to the definition, a cross stratified bed or layer is a sedimentation unit, as it has been deposited under essentially uniform conditions. The current that deposited cross-stratified layer has more or less maintained a uniform direction and velocity of flow. Cross-bedded layer deposited above the first one is another sedimentation unit, that indicates change in the flow conditions.

Thus, variations in velocity of depositing current give rise to different sedimentation units or beds. Such beds may not have distinct discontinuities but they are composed of several divisions, where each division is characterised by a set of sedimentary structures. Separation of different sedimentation units is by bedding planes. These can be said as planar, if bounding surfaces are parallel within the outcrop and therefore are laterally continuous. If the bounding surfaces converge or are irregular, the term lenticular is used, and if the bounding surfaces pinch and swell or disintegrate into discrete lenses, the terms wavy, slumpy, or nodular are used.

On the bases of regularity of sequence of beds, uniformity in thickness from bed to bed, lateral continuity and uniformity in thickness of individual beds, four classes of bedding have been identified, as:

(*i*) beds equal or subequal in thickness, laterally uniform and continuous,
(*ii*) beds unequal in thickness, laterally uniform and continuous,
(*iii*) beds unequal in thickness, laterally variable in thickness,
(*iv*) beds unequal in thickness, laterally variable and discontinuous.

As stated earlier, bed is generally characterised by a characteristic set of sedimentary structures. However, in case of structureless bed, it is termed as massive. Many beds, however, show the presence of internal laminations, which are parallel to the bounding surfaces of the bed. These laminations represent transitory or minor fluctuations in velocity of the depositing current.

In some cases, laminations with low angles. of about 1° or 2° or with higher angles between 10° to 35° or even with more angle have been recorded. Laminations with higher angles of inclination are termed as cross-stratification.

Ripple marks and cross-stratification

Important properties as the internal structure of a bed are cross-stratification and ripple marks. These are the characteristic features of a sandstone bed. However, even coarse and fine clastics exhibit these internal structures. Cross-stratification and ripple marks are two important features of a sedimentary rock, which are usually considered as two separate phenomena. Cross-stratification is considered as an internal feature of a bed, whereas ripple marks are considered as bedding plane structures. However both these features are related to each other. Cross-stratification is a result of the migration of the mega ripples or sand waves. Ripples on migration give rise to small scale stratification or ripple bedding.

Ripple marks is a small scale feature. Its parameters of measurements are wavelength, generally expressed in centimeter and its amplitude, more commonly measured in mm. In certain environments, however, formation of mega-ripples takes place, where wavelength of ripples may range from even a meter or more and amplitude may be in tens of cm.

As stated earlier, formation of these structures takes place due to flow of current. Such a flow in alluvial channel can be classified into lower flow regime and upper flow regime with a transition zone in between.

In lower flow regime, resistance to flow is large and sediment transport is generally small. In case of upper flow regime, resistance to flow is relatively small, sediment transport therefore, is relatively large. In lower flow regime, formation of the bed forms, such as; small ripple or mega ripple or of both may take place. In case of upper flow regime, however, formation of plane bed or antidunes may take place (Fig. 6.1).

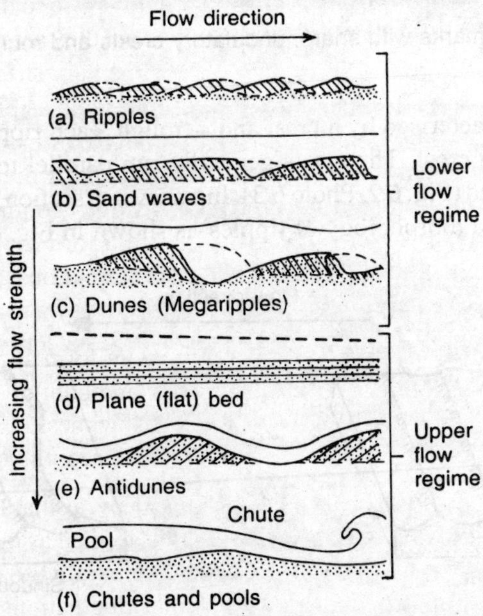

Fig. 6.1. Development of bed forms in upper and lower flow regimes.

Small ripples are characterised by gentle upstream slope and steep downstream slope, with ripple length usually less than 30 cm. In the initial stage of their formation, they have long, parallel crests, of small amplitude. With an increasing flow velocity, irregular and larger ripples are formed.

Mega ripples are more or less similar in shape to small ripples. Their length generally varies between 60 cm. to several meters. Their lengths can increase with an increase in depth of the channel. Such ripple marks commonly develop as undulations on a non-cohesive surface; even sometimes in muddy sediments. These are the result of an interaction between waves or currents and sediment surface.

Photo 6.2. Ripple marks with sharp, undulatory crests and rounded flattened troughs.

A ripple mark is characterised by a crest and a trough, each ripple crest has a trough that is shared by the adjoining crest. The crests generally run parallel to each other, which may be sharp, rounded or flattened (Fig. 6.2, Photo 6.3). In transverse section, ripples may be symmetrical or asymmetrical. General morphology of ripples is shown in Fig. 6.3.

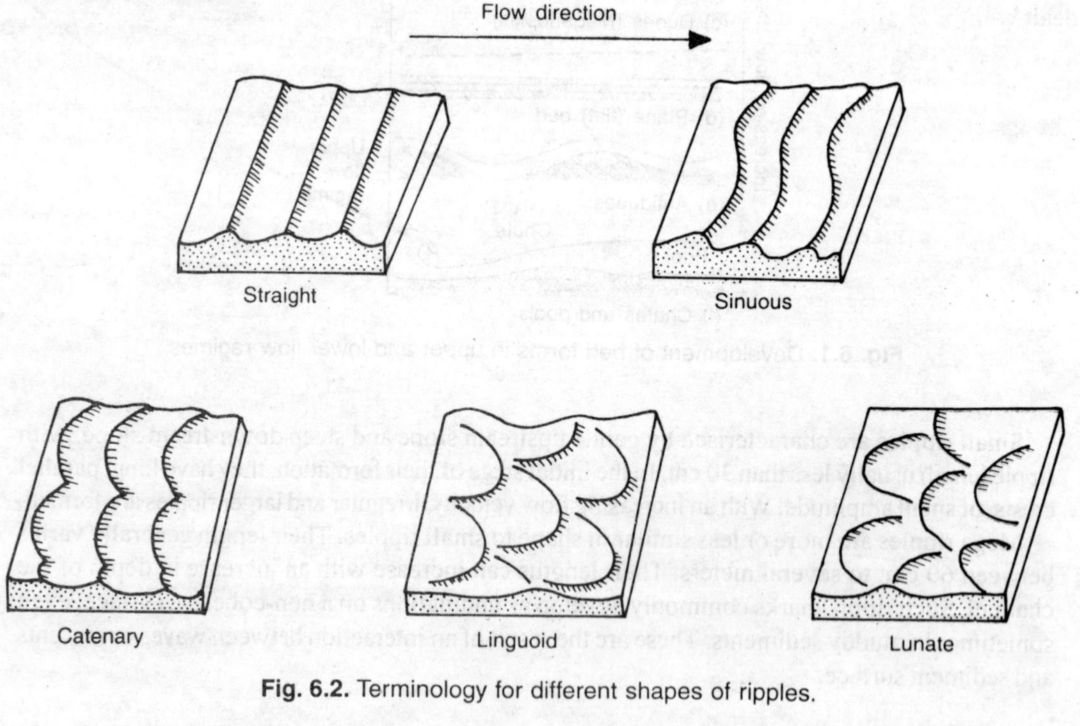

Fig. 6.2. Terminology for different shapes of ripples.

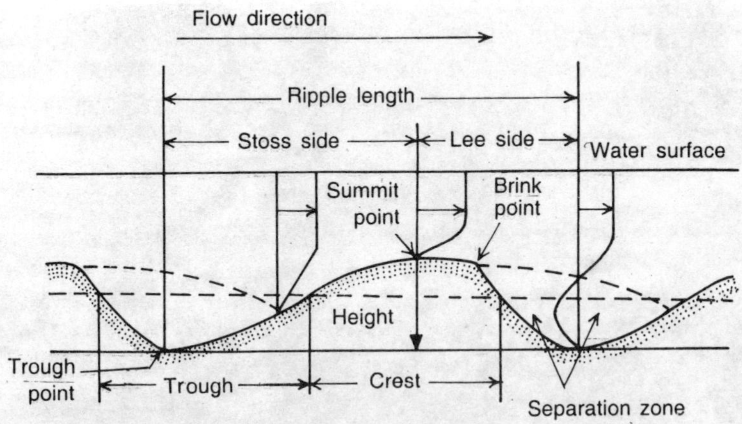

Fig. 6.3. Schematic cross-section of a ripple.

Various terms are used to describe ripple morphology Thus, as shown in figure, the terms are ripple length, ripple height, summit point, toe point, brink point, trough point, lee side, stoss side, slip face, bottom set, ripple index and symmetry index. Based on the nature of the crest of ripples, crests can be termed as straight, undulatory (Photo 6.3), linguoid, cuspate, lunate or rhomboid.

Origin of ripple marks is due to the action of current or wave or by the action of both (Photo 6.4). Ripple marks originate in different environments. However, identification of environment of deposition, on the basis of ripple geometry, is not conclusive. Reineck, *et al.* (1971) have dealt with in detail about the genetic classification of ripple marks.

Photo 6.3. Sharp, linear crested ripple marks.

Photo 6.4. Sandstone showing interference ripples.

Ripple Bedding

An important aspect of sand ripple is its internal structure. It is generally complex, and is referred to as cross lamination, which is relatively on a small scale. In the simplest form, it is cross-stratified layer, having thickness of about one cm. or more. If formation of ripple persists through time, superposition of one rippled bed above another takes place, and that gives rise to complex composite bed. Superposition of ripples may be in phases, so that it appears that ripples have not migrated. More common relation is superpositions of ripples with progressive migration of the crest, which gives an appearance that ripples climb on the stoss-side of the underlying ripple.

Superposition of several sets of ripples, entirely out of phase with each other, gives rise to lens like structure, referred to as flaser bedding (Fig. 6.4).

An irregular stacking of ripple bedding results into complex internal lamination. It is due to rippling but well organised persistant stacking pattern is absent. Such a structure is termed as wavy bedding. Presence of mud makes the form of ripple bedded unit much evident. The intercalated mud may be present either in the form of small lenses or flasers, where isolated accumulation of mud in ripple troughs takes place. This gives rise to flaser bedding. If small lenses of mud meet each other formation of wavy bedding takes place. In case, if mud dominates and ripple bedded units are isolated in a mud matrix, it is termed as, lenticular bedding or starved ripples.

Another internal structure of sandstone, which is related to ripple bedding is 'rib and furrow' structure. On a bedding plane, this structure appears as small, transverse, crescent-shaped marking. These markings occur in sets which are confined to long narrow furrows, separated from one another by narrow and discontinuous ribs. The longitudinal furrows are generally parallel to each other and are in the direction of current. They are about a meter in length and a few cm. in width. The transverse markings are arcuate, their convex side is in the current direction, while bisectrix is parallel to the current.

The rib and furrow appears as the bedding plane expression of small scale trough cross-stratification, formed due to migration of linguoid ripples.

Fig. 6.4. Flaser structure.

Cross-Stratification

Of the different sedimentary structures, cross-stratification or cross-lamination has received wide attention in sedimentological studies and research. It is because that its morphological parameters can easily be quantified, which further lead to palaeocurrent and basin analysis studies.

Cross stratification is commonly described by following terminology. When a single cross-stratified unit is bounded by bedding plane, it is termed as set. If a group of similar sets are not separated from each other and there is an absence of major bounding plane, it is termed as a coset (Fig. 6.5). If a set having upper surface has preserved the shape of the original bed form, it is termed as form set. Cross-stratification may also be classified on the bases of scale of sets, shape and attitude of cross-stratified units and on the nature of lower and upper bounding surfaces. Cross-stratified units or sets can be termed as tabular, if bounded by essentially plane and parallel surfaces If it is bounded by plane but non-parallel surfaces and trough sets, where lower bounding surface is scoop shaped, it is wedge-shaped cross-stratification.

Other synonymous terms applied to cross-stratification or cross-lamination are cross-bedding, diagonal bedding, current bedding or false bedding. Cross stratification, however, is generally restricted to internal bedding, termed as foreset bedding. It is inclined to a principal surface of deposition within a single sedimentation unit. The term cross-stratification is used if thickness of foreset layers is greater than one cm. and if it is less, the term cross-lamination is used.

A simple cross-stratified layer is described by its scale, inclination and azimuth. Scale is the thickness of the cross-stratified unit, which varies between one and two cm. Cross-stratified sets are generally less than one meter thick. Inclination indicates dip of the foreset laminae. It is the dihedral angle between the foreset plane or a plane tangential to the foreset at the place of its greatest inclination. An average angle of inclination of cross-stratified units lies between 15° and 20°. The azimuth is the horizontal angle between the meridian and the horizontal projection of the

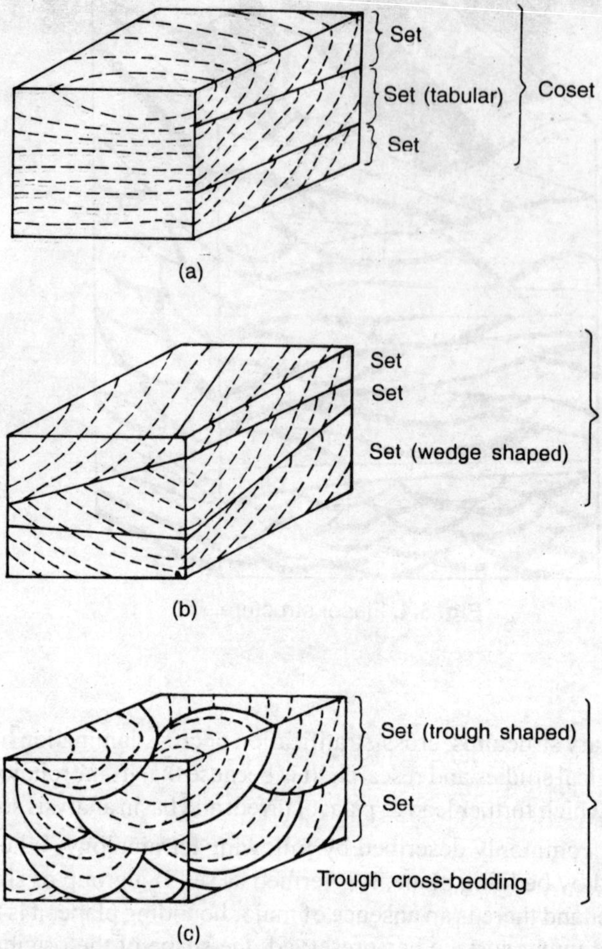

Fig. 6.5. Terminology for class-bedded strata (after McKee & Weir, 1953).

dip line of the foreset, which is in the down current direction. In case of trough shaped cross-stratified units, traces of foresets are curved and concave in down current direction. The pattern of bedding planes in small scale cross-stratification is termed as 'rib and furrow' or micro cross-lamination.Cross-stratification is a result of the migration of sand waves. Size of sand waves determines scale of cross-stratification. Migration of dunes gives rise to large scale cross-stratification, while ripple migration produces small scale cross-stratification. Regular linear ripples or sand waves give rise to simple, planar-tabular cross-stratification, (Photo 6.5) whereas linguoid wave forms result into trough cross-stratification.

Presence of cross-stratification in lithological unit shows preferred orientation. In case of alluvial deposits, mean direction of current is downslope, but azimuth is in opposite direction. This suggests ebb-flood tidal current and such a cross-stratification is described as herring-bone structure. Though no particular scale of cross-stratified unit is considered as a characteristic of any particular environment, large scale cross-stratification is generally considered to be aeolian or marine rather than of fluvial origin. However, scale of cross-bedding in subaqueous deposits is related to water depth.

Photo 6.5. Cross bedding in sandstone.

Graded Bedding

It is a common feature of some sedimentary sequences. The presence of structure is order of superposition, and is particulerly useful in complexely folded and overturned strata. It is considered to be the characteristic feature of turbidity current deposits, formed in aqueous conditions, at greater depth.

Each bed in graded bedding is a sedimentation unit, and therefore graded beds are sedimentation units. It is characterised by gradation in grain size, where the lowermost unit, is marked by coarse grained sediments, which gradually become finer and finer to the top of the unit. Graded beds are the deposits of a waning current system. Thickness of graded beds varies from a cm. or less to one or more meters. It generally develops in sandstone. It has been recorded that more the thickness of graded bed, relatively coarser material is present at the base. Thickness of graded beds as observed in one unit, displays a log normal distribution.

In majority of the graded units, ideal and normal sequence of graded structure is encountered, though variations in their pattern are also observed. An ideal cycle of graded structure is termed as the 'Bouma cycle'. As suggested by Bouma, an ideal sequence of graded bedding units consists of the following, five major divisions or intervals:

(*a*) lowest unit of the graded bed exhibits marked grading and it is generally the thickest part of the sequence. In case, well sorted sand constitutes the basal part of the sequence, graded structure is rather indistinct, and represents the upper flow regime;

(*b*) in case, the thickest and the basal and graded structure is followed by an interval of parallel laminated bed, it also represents upper regime;

(*c*) an interval of laminated bed in turn gives rise to ripple cross-lamination, which results in lower flow regime conditions;

(*d*) according to Bouma, ripple cross-laminated layer is followed by an upper interval of indistinctly laminated sandy or silty pelitic layer; formed mainly because of superposition, such as interval layer. However, it is found to be poorly developed.
(*e*) the topmost interval is either shale or slate, i.e. of the pelitic material.

It has been observed that in many cases, a complete, ideal sequence is not developed. It is either the lower most or the top most layer that is found to be truncated. As per the observations made by Bouma, if any layer in the sequence is found to be truncated, development of other layers or intervals in the sequence is generally complete. An incomplete development of sequence can be attributed to sudden weakening or waning current, depositing initially, the coarser sand.

It has also been recorded that there is a lateral change in character of graded beds. It is marked by a decrease in thickness alongwith the decrease in grain size. It, therefore, follows that decrease in thickness of graded beds, associated with decrease in grain size results in the down current direction. An ideal sequence of graded bed, therefore, is apt to develop near the source while in more distal beds, truncation of bottom beds may take place. Development of the ideal sequence therefore, indicates proximity to the source. Away from the source, there is gradual truncation of lower beds, associated with decrease in grain size.

Graded bedding and cross bedding are mutually exclusive features.These do not develop in the same sedimentary sequence. However, small scale cross bedding and small scale ripple cross-lamination develop in an ideal graded bed but large scale cross bedding does not develop in graded bed.

Graded bedding is primarily a characteristic feature of sandstone and also that of graywackes of older geological sequences.

Formation of majority of the graded bed sequences is ascribed to turbidity current. Such a feature has been produced experimentally by turbidity flows (Kuenen and Migliorini, 1950; Kuenen and Menard, 1952).

The graded beds may also originate by other ways. Presence of varved beds and sands in periglacial lakes suggest a seasonal influx of sediments, controlled by seasonal melting of a glacier that is responsible for the formation of a graded bed. However, this view for the origin of a graded bed is debatable as the formation of graded beds is assumed to be seasonal and it implies high rates of sedimentation.

According to Kuenen (1953), several hundred to several thousand years get elapsed between the formation of the graded bed and the next succeeding one. Graded beds, however, record relatively short lived events. Isolated or sporadic graded beds may originate either due to volcanic eruption or heavy floods or storms. Most of the repetitive massive graded beds are the products of turbidity current flows.

Bedding Plane Markings and Structures

When beds separate along bedding planes, surfaces that are produced, are commonly marked by various structures.

Such features form on the surface of accumulating sediments. On these bedding plane surfaces, most of the features are preserved as casts on the underlying surface or the sole of the overlying bed. Features, such as; rain prints, mudcracks, flutes and grooves are generally preserved as casts on the underlying surface of the overlying bed; whereas some of the features appear either on the underside or topside of a bed, e.g. ripple marks appear as casts and also as normal structures on the bedding planes. Similarly, feature of parting lineation also appears. The features which characteristically develop only on mudstone appear as sole marks, whereas the features that develop on sandstone appear on the bedding plane surface and also on the underside surface of the bed. In the following paragraphs, features which can develop on mudstone have been described.

Sole Marks

These are the bedding plane features, which develop on the undersurface of sandstone and also of limestone, if it overlies shale. These features are formed due to filling of depressions on the mud surface, over which sandstone bed has been deposited. These markings originate by current action, by deformation due to loading and also by the action of organisms. The features which are produced due to current can be classified into two groups; as those which are produced due to scouring action of current and those formed by the action of debris carried by the current. The features classified in the earlier category are termed as scour marks, while those in the latter are termed as tool marks.

Scour Marks

Scouring action of current may give rise to flutes, which later become filled with sand and get welded to the overlying sand bed. These are termed as `flute casts'. These casts, therefore, appear as raised structures on the base of the overlying sand bed. The cast is slightly elevated, elongated, having mound-like form, with upcurrent tapering end. Upcurrent side of the flute merges with the bedding plane. Size, shape and arrangement of the flute casts vary from place to place. Shape of these is generally triangular or deltoid. Its prominant upcurrent end is either twisted or displays hook-like appearance. The flute casts generally occur in swarms and are widely spaced or may cover most of the sole area(underlying area) of the sand bed.

The flute is found to be filled by sand, generally coarser than the sand bed of which flute casts form a part. Less regularly developed flute casts may have an appearance of load casts. The flute, however, is cut or eroded and transects laminations of the underlying bed. In constrast, laminations around load casts are deformed, but are not transected by them.

Origin of flutes is ascribed to the action of eddy currents. Under favourable conditions, when eddy currents are produced, they scour the underlying mud surface. Size of the flutes depends on coarseness of the material and strength of the current. These are, therefore, of help in paleocurrent studies. Flute casts most prominantly are developed on the underside of turbidite sandstone bed and therefore, are the characteristic of flysch facies.

Another related structure to scouring action of current is the current crescent. It results due to scouring around an obstacle, lying on a sandstone bed. The scouring is greatest on the upcurrent side. It also extends on either side in the downcurrent direction. Object of obstruction is generally weathered later, which leaves behind a hole, that is partially surrounded by a crescentic, ridge-like structure.

Tool Marks

Current also carries with it sedimentary particles, of various sizes, which travel on and also across the underlying mud surface. These particles or tools travel either by rolling or by saltation mode, i.e. intermittently touching or hitting the surface of the underlying bed, leaving its mark on the surface. These marks may be preserved as positive features on the base of the underlying sandstone, which are collectively termed as tool marks (Fig. 6.6).

One of the prominent tool marks is groove cast. It is a raised, rectilinear feature, having sharp to rounded crests It develops on the base of some sandstone beds. It may be multiple in number and at

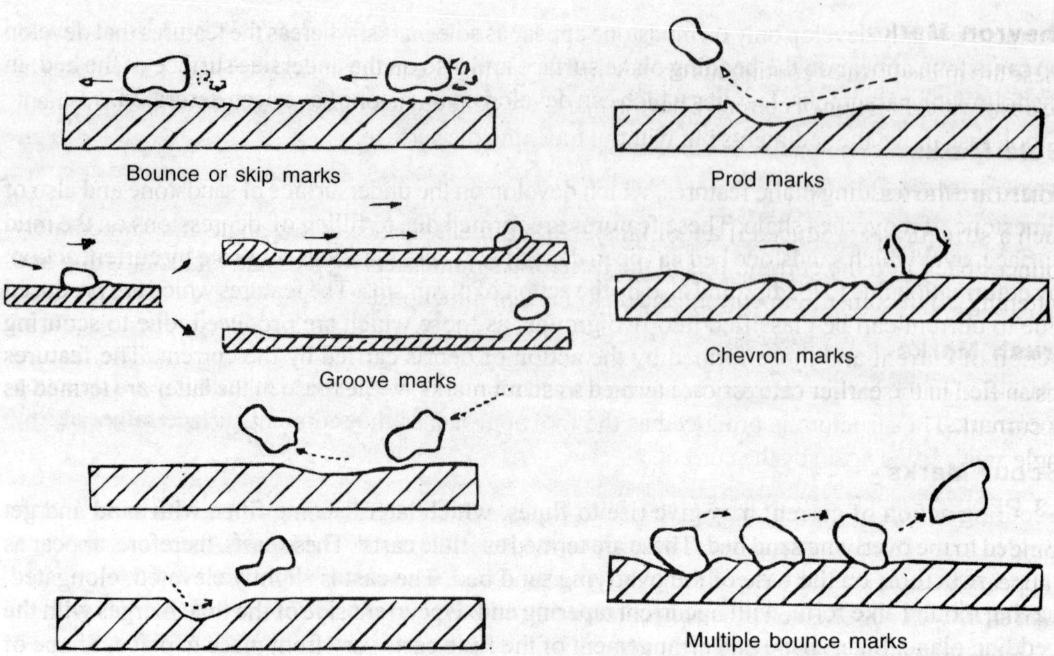

Fig. 6.6. Tool marks.

times exhibit development of second order set of grooves or ridges. These are characterised by smaller, slightly divergent grooves. Such second order grooves are arranged more or less in a symmetrical manner on both the sides of a principal groove. Various objects moved by current on a mud surface produce furrows, which are later filled and groove casts are formed. These are also termed as drag marks and drag casts.

Groove casts generally occur in more than one set. One of the sets is commonly seen on the same surface, while the second set intersects the first set at an acute angle. Azimuth of grooves in a particular set is oriented commonly in one direction. Groove casts and flute casts occur very rarely together, but like flute casts, they are also characteristic feature of turbidite sandstone bed and therefore, the common sole marks of the flysch facies.

Groove casts are useful indicators for palaeocurrent direction. However, they are used alongwith other structures, because grooves indicate only the azimuth but not direction.

Apart from groove casts, other features also develop as tool mark on sandstone bed. These features are produced as the tools move either by rolling on the surface of the bed or by touching the surface intermittently. The features developed by the latter action are brush, bounce and prod casts. Brush casts develop due to tool impact but which is accidental and therefore, these are not regularly repeated. These appear as a small mound of the material that is pushed up ahead of the obstructing objects. Bounce casts are produced because of saltating mode of movement of the tools or objects. These are small, slightly raised features, and which occur more or less with regular intervals. The prod casts originate due to penetration of the mud bottom by an object, which is rotated forward in the stage of penetration and later lifted out of the mud. Prod casts, therefore, have more or less a similar appearance as that of groove casts. Rolling objects give rise to rolling marks, which commonly occur in the flysch sequence. It is a feature that appears like tread on an automobile tyre.

Chevron Marks

These are in the form of continuous open V marks, giving rise to linear ridges. The V form closes in the downward direction of current. Chevron marks are believed to originate because of the movement of tool, just above the sediments but without touching the surface.

Bounce Marks

Such a structure is produced if a tool falls on a sediment surface at low angle and immediately bounces back into the current. It is in the form of a symmetrical depression and is tapering and flattening off in both, the upcurrent and downcurrent directions.

Brush Marks

It is an elongated shallow depression, having a small rounded circular ridge in the downward direction of current. The structure is produced as the tool approaches the sediment surface rather at low angle and is lifted again by the current.

6.3 PENECONTEMPORANEOUS DEFORMATION STRUCTURES

Some of the sedimentary structures, when observed, indicate that there has been some deformation. It appears to have formed either during the process of sediment deposition or immediately after it. Such structures are referred to as penecontemporaneous deformation structures.

The deformation structures appear to have formed due to combination of more than one of the following mechanisms:

(*i*) during the process of sediment deposition, if sediments with low density are overlain by sediments with relatively higher density, reverse density gradient is established due to gravity;
(*ii*) liquefaction of sediments;
(*iii*) sediment deposition on sloping surface results in sliding and slumping due to gravity;
(*iv*) a flow that moves recently deposited sediments exerts shearing stresses.

Accordingly, various penecontemporaneous deformation structures are formed.

Structures that are formed due to deposition of sand over hydroplastic mud layer are termed as, load structures. These are sole markings, which are generally preserved on the lower side of the sand layer. In this structure, it is a pattern of bulbous protuberances, separated by relatively narrow ridges. The protruding material is generally sand that sink into the fine-sized material, which was plastic at the time of deposition. Such a plastic layer gets distorted and bends downward below the lode protuberances. Due to loading, the earlier formed structures, such as; flutes or grooves also are modified and distorted.

Thus, in case of ripples, load structures are produced due to differential compaction. In such a case, ripples sink into the underlying mud. However, unequal loading not necessarily results into vertical movement, but also because of lateral movement due to flowage (Photo 6.6).

Another form of such a structure is observed, which appears as pseudonodules. It is a sand or silt layer, that attains a lower convex surface on distortion and is completely engulfed in mud. Ball and pillow is a similar structure (Photo 6.7) but it differs mainly in smaller amounts of mud or shale. In this structure, layer of sand is broken into number of 'pillows' having rounded bottoms (Fig. 6.7). These pillows may be connected or at times are totally detached, and are floating in a matrix of mud. Upper part of the bed, however, remains rather undisturbed. Formation

Photo 6.6. Convolute laminations in sandstone.

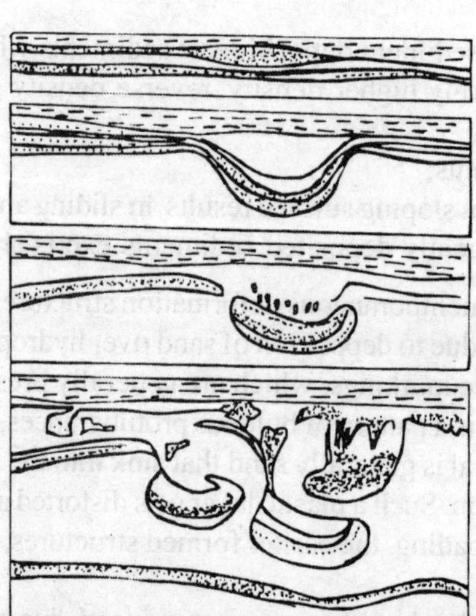

Fig. 6.7. Ball and pillow structure.

of the structure is ascribed to the partial failure of the sand layer. The pillows may exhibit some structure, as laminations or are devoid of any structure (Photo 6.6). Presence of distorted laminae suggests incomplete liquefaction of the sand layer during its formation.

Origin of these deformation structures has also been ascribed to slumping of strata. However, slump structures are difficult to identify in the field and cannot be distinguished from load or liquefaction structures. However, slump structures can be identified by distorted and overturned strata, resulting into thickening and thinning; as sediments behaved plastically at the time of deformation. Slump structures that originate under the influence of gravity are generally associated, with rapid sedimentation. Slump structures in associated with brecciated horizon and small scale faults have also been reported .

Another distinctive type of deformation structure is convolute bedding (Photo 6.7). It develops generally into fine sand layer. In this case, parallel laminae or foreset laminae of ripple bedding is thrown into convolution due to deformation. In spite of intensive deformation, each laminae in a structure is remarkably continuous. Such a deformed bed exhibits a complex pattern of broad synclines and sharp anticlines. Such folds are generally overturned, but within the bed itself. It has been inferred that formation of such structures results because of plastic deformation at the time of rapid deposition from suspension. This suggests that shear stresses responsible for deformation have been caused because of the current.

Dish Structure

These structures develop parallel to bedding, in sandy beds. In a dish, upper layers are of coarse sand, having clay rich laminae. Concavity of a structure increases upward, with a corresponding decrease in width. This structure develops due to escape of water and is formed at the time of consolidation and dewatering of rapidly depositing sediments. It is, therefore, included in penecontemporaneous deformation structures. Environments of delta fan and alluvial fans are favourable for development of such a structure.

Photo 6.7. Ball and pillow structure.

6.4 BIOGENIC STRUCTURES

It is observed that sediments after deposition are reworked by organisms. Resulting structures are termed as biogenic structures, trace fossils or ichnofossils. Biogenic structures, which are formed at the sediment surface are termed as exogenetic, while those which are formed below the sediment surface are termed as endogenetic. However, a distinction between the two is rather difficult, due to burrowing activity of the organisms.

A significant biogenic structure is 'stromatolites'. It is a laminated structure and is composed of sand-silt and clay sized sediment particulates. These are trapped and bound by detrital sedimentary particles by an algal mat. It varies from flat laminations to small mound like forms of varying convexity and size to columnar structures (Photo 6.8). The external form and size are more or less hemispherical. The concentrically structured bodies that appear like concretions and attached to stromatolites are termed as oncoliths.

The structure like stromatolite but lacking internal laminations is termed as thrombolites.

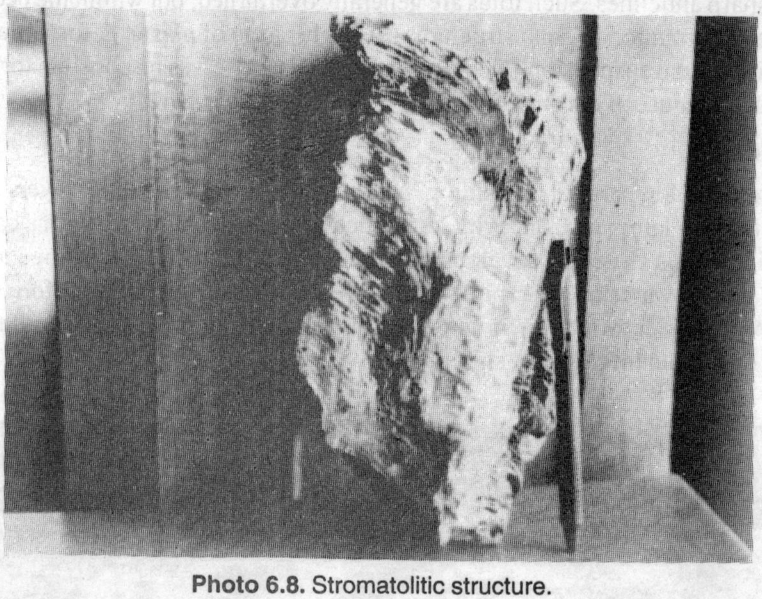

Photo 6.8. Stromatolitic structure.

There are different views about the genesis of stromatolites According to the earlier view, stromatolitic structure is formed due to secretion by specific organism. However, this view was not accepted. Later, it has been proposed that the algal mats are responsible for the formation of such a structure, wherein a variety of green and blue-green algae played a significant role. It has also been regarded that various stromatolitic forms are environment-dependent rather than generic-dependent. Stromatolites formed by algae have fine laminated, fragmental texture. Various forms of stromatolitic structures are diagramatically presented in Fig. 6.8. These stromatolitic structures develop in limestone and have been reported from Precambrian to Recent strata.

Other biogenic structures include tracks, trails and burrows. Study and analysis of these structures reveal the history of sedimentation.

These biogenic structures result due to particular activities of organisms, like dwelling habit or

Fig. 6.8. Different types of stromatolitic forms.

feeding habit; which in turn indicate the environment in which organisms had lived. The information about the sedimentary facies and water depth can also be inferred.

Trace fossils provide information about the rate of sedimentation and the stratigraphic order of superposition, in case where strata are steeply dipping or overturned.

Seilacher (1964, in Pettijohn, 1974) has identified five functional classes of trace fossils on the basis of animal behaviour, which are as follows:

1. *Resting marks (Ruhrspuren)*: resting marks are the shallow marks, made by organisms, which rest on the bottom.
2. *Crawling trails (Kreichspuren)*: crawling trails are produced by mobile organisms, when they move on the sediment surface.
3. *Residence or shelter structures (Wohnhauten)*: residence or shelter structures are because of mobile or semiattached organisms, generally to protect from scouring action by current. These are more or less permanent structures.
4. *Feeding structures (Fressbauten)*: feeding structures are generally the burrows having a radial pattern, which are made by sessile organisms.
5. *Grazing trails (Weidespuren)*: grazing trails are generally sinuous trails or burrows made by mud eating organisms, at the sediment-water interface.

These biogenic structures can also be classified on the basis of their relation to bedding, their geometrical pattern and ornamentation or internal structure of organisms.

There are also other biogenic structures, which can be observed in section perpendicular to the bedding. These are simple tubes or U-shaped tubes. Burrows can also be observed, which may be simple or branched in shape. The burrows are later filled by the different material from that of the host sediments. Burrows also ridge the sediment-water interface. Burrows are generally shallow or may extend downward a long distance and largely may be confined in a horizontal plane at the interface, commonly within a given bed or between beds. Complete absence of burrows or other trace fossil is a strong indication of euxinic conditions prevalent at the time of sediment deposition.

Some of the trace fossils may also provide information about the rate of sedimentation, if the organisms have produced resting tracks just below the sediment interface. However, if the same is covered by a sudden influx of sediments, the same organism burrows the new surface and produces another set of resting tracks. Presence of number of burrows at sediment interface is an indication of long period of non-deposition.

Number of burrowing organisms commonly eat sediments, generally mud, for its organic content and give rise to feeding burrows. These organisms then excrete mud, in the form of fecal pellets.

6.5 PALAEOCURRENT ANALYSIS

In palaeocurrent analysis, study of the directional and other primary properties of the sedimentary rocks is carried out to reconstruct the current system, to establish the direction of palaeoslope, to establish the relation between facies boundaries, to delineate the sediment supply and finally to construct the basin configuration.

Palaeocurrent study contributes to the fields like sedimentary petrology, palaeoecology, geophysics, sedimentary tectonics, regional and local mineral exploration. The information regarding grain size, roundness, bedding characteristics, depositional strike and symmetry concepts in sedimentary fabrics within the depositional basin can also be gathered from these studies.

Palaeocurrent analysis can help in predicting local and regional sedimentary trends in exploration of ground water and petroleum. One of the palaeocurrent features, i.e. depositional fabric, is of help in interpreting geophysical properties like permeability. Mapping of the directional structures also provide useful information about sediment tectonics that operated during deposition. Palaeocurrent directions also help in palaeoecological studies as currents supply nutrients, may remove waste products and may also disperse and orient organisms.

In palaeocurrent analysis, measurement and orientation of the significant primary sedimentary structures, exposed in the field, are carried out. For the purpose, measurements at each sample point are made and regional palaeocurrent map is prepared. By combining palaeocurrent map with that of facies analysis, palaeogeography and environment of deposition are interpreted.

With the help of field observations, laboratory studies are carried out to prepare palaeocurrent map and inferences, stated above are drawn.

The palaeocurrent studies are based on four major properties of sedimentary rocks namely, attributes, scalars, directional or vectors and tensors.

Attributes indicate the property which is specified by its presence or absence; e.g. presence of heavy minerals. If mapped systematically, proportion of heavy minerals varies according to current direction.

The property of scalars is specified by magnitude, e.g. mean grain size of sedimentary rock or sand-shale ratio. Such properties bear relations with current direction.

Directional or vectorial property indicates either a line of movement or a direction of movement and are specified by azimuth. e.g. ripple marks, cross bedding, sedimentary fabric, etc.

Tensors are specified by some geophysical properties, which depend on permeability, electrical resistivity, thermal conductivity, etc.

With the help of four different types of properties, data is collected. It is commonly presented by rose diagramme or rosettes. These rosettes are either unimodal or bimodal or polymodal. More information can be gathered by combining palaeocurrent data with that of the facies.

Average or mean current direction and also the magnitude of paleocurrent can be obtained by statistical analysis of the data.

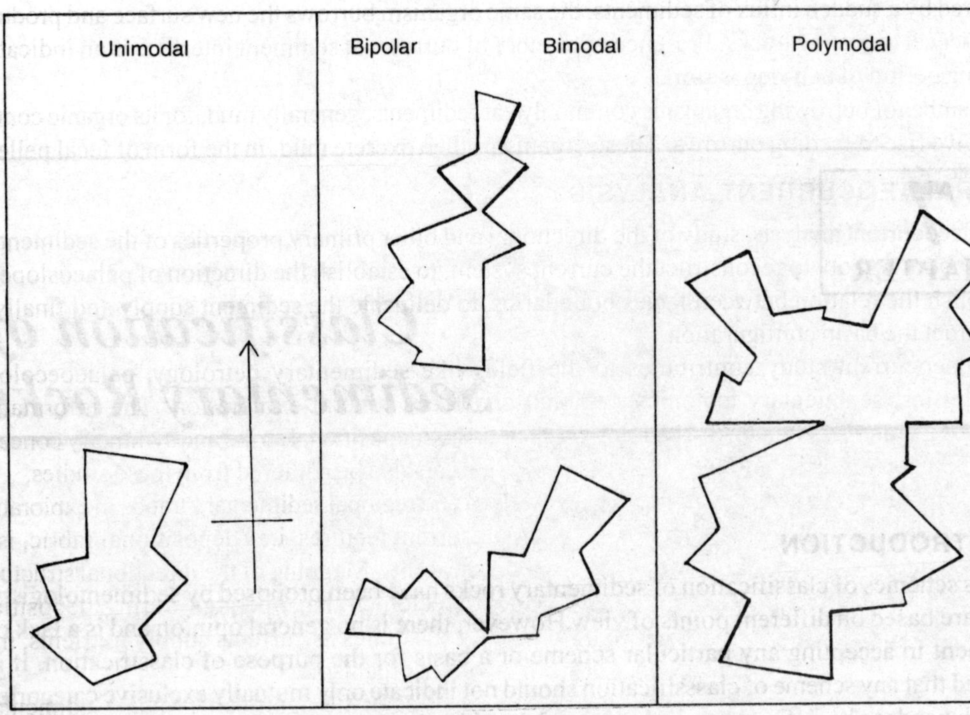

Fig. 6.9. Genera patterns of palaeocurrent direction.

The palaeocurrent data interpreted from field observations and that from the laboratory studies is presented graphically (Fig. 6.9) and also analysed statistically. In combination with the information about the sedimentary facies, it helps to prepare the regional palaeocurrent map of the area. Interpretation of such maps is then carried out, according to the purpose of analysis.

7
CHAPTER

Classification of Sedimentary Rocks

7.1 INTRODUCTION

Various schemes of classification of sedimentary rocks have been proposed by sedimentologists, which are based on different points of view.However, there is no general opinion and is a lack of agreement in accepting any particular scheme or a basis for the purpose of classification. It is expected that any scheme of classsification should not indicate only mutually exclusive categories but each needs to be defined precisely to stand for objective test of determination.

Commonly accepted principles for classification of sedimentary rocks have been discussed in the following paragraphs;

A primary and fundamental aspect of calssification is descriptive, while more advanced levels are genetic classification.

Descriptive schemes of classification have the following three aspects:

(*a*) Certain·fundamental properties of sediments are considered as more basic than others. In some classifications, size of sedimentary particles and mineralogical composition have been regarded as more important properties than other characteristics of sediments.

(*b*) Different schemes of classification have been devised, However, each one is based on a single fundamental property, such as; size or shape or packing, etc. and no consideration is given to mineralogical composition.

(*c*) Sedimentary rocks have been recognised by grouping them as measurements of various parameters. Thus, a bivariate plot is made on the basis of two parameters. Scatter diagram is searched for point, which shows grouping or cluster of the points and those which are scattered from such group. Separating such groups by a dividing line, demarcates the regions of low and high density with repsect to the cluster of points. This method enables descriptive classification of sedimentary rocks.

Sedimentary rocks have also been classified on the basis of genesis. Schemes of genetic classification stand at more advanced level than the descriptive classification. However, many difficulties are encountered in genetic classification, as most of the sedimentary rocks have complex origin and therefore, it becomes difficult to single out an important mode of genesis. However, schemes of genetic classification of sedimentary rocks have been grouped on the basis of following parameters:

1. Those derived from same source;
2. Those having a similar history of transportation and erosion;
3. Those which are deposited in the same environment; and
4. Those depostied by the same physical or chemical processes.

While classifying a sedimentary rock, both the aspects, i.e. genetic and descriptive are considered; so that a rock 'term' indicates physical and mineralogical characteristics and also about its origin. However, such an ideal classification has rarely been attempted.

The following broader schemes of classification of sedimentary rocks are based on genetic aspects.

7.2 CLASSIFYING SEDIMENTARY ROCKS

Fundamental classification of sedimentary rocks is given by Folk (1974). According to him, sediments primarily consist of three components, which may be mixed in all or different proportions. These are (*i*) Terrigenous components, (*ii*) Allochemical components and (*iii*) Orthochemical components.

(*i*) Terrigenous components are those that are derived from erosion of land areas, outside the basin of deposition and transported into the basin, as solids; e.g. quartz, feldspar, heavy minerals, etc.

(*ii*) Allochemical constituents are those that are precipitated from solution within the basin of deposition, but are rather 'abnormal' chemical precipitates, because in general, they are later transported as solids within the basin, e.g. oolites, pisolites, faecal pellets, etc.

(*iii*) Orthochemical constituents are the 'normal' chemical precipitates in the general sense. They are formed chemically within the basin and show little or no evidence of significant transportation or aggregation into more complex substances.

Classes (*i*) and (*ii*) are collectively referred to as 'Fragmental', while classes (*ii*) and (*iii*) are collectively referred to as "chemical" constituents. The terms 'detrital' and 'clastic' are used as synonyms for 'terrigenous'.

Based on the relative proportions of three components, sedimentary rocks have been classified into five different groups. These are presented by a triangular diagram (Fig. 7.1).

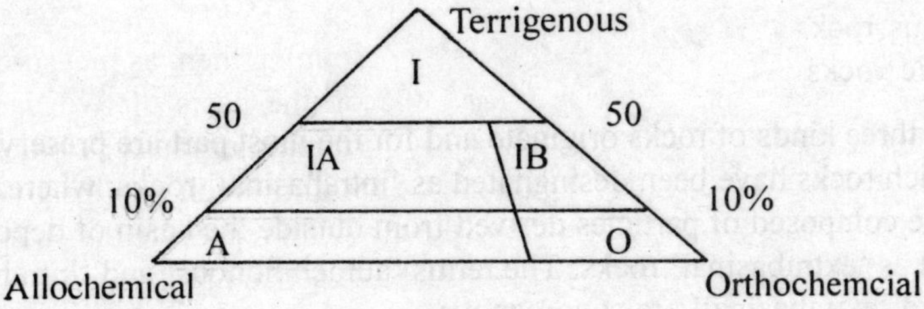

Fig. 7.1

I. Terrigenous rocks, e.g. most mud rocks, sandstones, and conglomerates comprise nearly 65% to 75% of the stratigraphic section;

IA. Impure allochemical rocks, e.g. fossiliferous shales; sandy fossiliferous and oolitic limestone comprise nearly 10% to 15% of the stratigraphic section;
IB. Impure orthochemical rocks, e.g. clayey microcrystalline limestone comprise 2% to 5% of the stratigraphic section,
O. Orthochemical rocks, e.g. microcrystalline limestone or dolomite, anhydrite, chert comprise 2% to 8% of the stratigraphic section,
A. Allochemical rocks, e.g. fossiliferous, oolitic, pellet or intraclastic limestone or dolomite comprise 8% to 15% of the stratigraphic section.

Another scheme of classification is based on the origin and derivation of the constituents. Accordingly, the sedimentary rocks are broadly grouped into two, as; Intrabasinal or Autochthonous and Extrabasinal or Allochthonous rocks. These are discussed in the following paragraphs;

Group-I: Autochthonous Sediments

Rocks included in this group are formed in situ and have undergone least transportation;

a Residual Deposits : Laterite, Bauxite, etc.
b Chemical precipitates : Evaporites, gypsum, rock salt, etc.
c Organic Depostis : Coal, limestone, dolomite, etc.

Group-II: Allochthonous Sediments

Rocks included in this group are clastic in nature and have undergone different degree of transportation.

d Terrigenous Deposits : Conglomerates, siliciclastic sands, clays, etc.
e. Pyroclastic Deposits : Agglomerates and volcaniclastic sands, ash, tuffs, etc.

Friedman and Sanders (1978), on the same principles, have proposed five major kinds of rocks, as follows;

Intrabasinal rocks

1 Carbonate rocks : limestone and dolomite
2 Authigenic rocks : evaporites, cherts, iron rich rocks
3 Carbonaceous rocks : coal.

Extrabasinal rocks

4 Terrigenous rocks
5 Pyroclastic rocks

Of these, first three kinds of rocks originate and for the most part are preserved in the basin of deposition. Such rocks have been desingnated as 'intrabasinal' rocks; whereas, the rest two kinds of rocks are composed of particles derived from outside the basin of deposition and are, therefore, termed as 'extrabasinal' rocks. The terms 'autochthonous' and 'intrabasinal'; while, allochthonous and 'extrabasinal' are synonymous.

7.3 CLASSIFYING EXTRABASINAL OR ALLOCHTHONOUS ROCKS

Extrabasinal rocks include terrigenous rocks. A term terrigenous is derived from the Latin word 'terra', meaning land. These rocks are derived as sediments due to weathering and erosion of land masses, Such terrigenous particles are transported to the basin of deposition, and further compoaction and lithification give rise to terrigenous rocks, These are also termed as clastic rocks.

On the basis of the particle size, three main kinds of terrigenous rocks are identifed as; 1. shales, 2. sandstone and 3. conglomerates and breccias. According to the Grabau's system of classification, these are respectively termed as lutite, arenite and rudite.

Each scheme of classification is described separately in the following paragraphs;

Shales: Three major bases for naming these rocks have been proposed: which are a. particle size, b. proportion of clay minerals and c. fissility; which is a property of splitting into parallel thin bands.

Shales are terrigenous clastic rocks, composed of silt-and clay-sized particles. Megascopic inspection of shales indicates that majority of rocks termed as shales are really siltstones and these rocks contain particles of silt rather than of clay size. The distinction is, therefore, made between the dominantly clay shales from the dominantly silt shales. Identification of such a difference is significant to distinguish the transition between bottomsets and foresets in fine grained sediments of delta.

Proportion of clay minerals in shales varies widely. It is commonly stated that shales are composed almost entirelyof clay minerals, However, quartz may constitute from 25% to 50% in proportion; while alongwith feldspar, quartz and felspar; whereas, in fine grained shales, these are less but have greater proportions of clay minerals. Fine grained terrigenous sedimentary rocks have abundant proportions of quartz and feldspar, which are generally of silt size, Such rocks due to the abundance of these particles closely resemble to sandstone, However, these are generally referred to as siltstone, and lacks fissility, In contrast, fine grained shales contain more abundant clay minerals, which are oriented parallel to the bedding, but not necessarily imparting fissility to the shale.

Fissility is a characteristic of the depositional fabric. A shale is fissile and in its depositional environment, burrowing organisms are absent, According to the commonly assepted definition, a shale must exhibit fissility, and includes all the fine-grained sedimentary rocks, that contain substantial amounts of clay minerals, and fissility may or may not be present,

The terms mudstone, claystone and argillite are also commonly applied for shales, lacking fissility. Argillite is more indurated shale than claystone and mudstone.

Sandstone Classification

Various schemes of comprehensive and systematic classification of sandstone have considered 'sand-sized' grains in 'light' mineral fractions. In all the schemes, equilateral, triangular classification, having the three mineral species at the apices has been suggested, The apices of the triangle are the mineral species or rock fragments and furher subdivisions have been arbitrarily suggested by different authors, Each subdivision of the triangle has a different term. This has, therefore, yielded a variety of names to 'sandstone'.

Various schemes of classification of sandstone, based on 'light' mineral species and rock fragments have been presented in Fig. 7.2, that describes different triangular schemes of classification.

Practical advantage of such schemes of classification is that all the three mineral species of the apices can easily be distinguished, various schemes of classification can then be confirmed by preparing and examining representative microsection under petrological microscope.

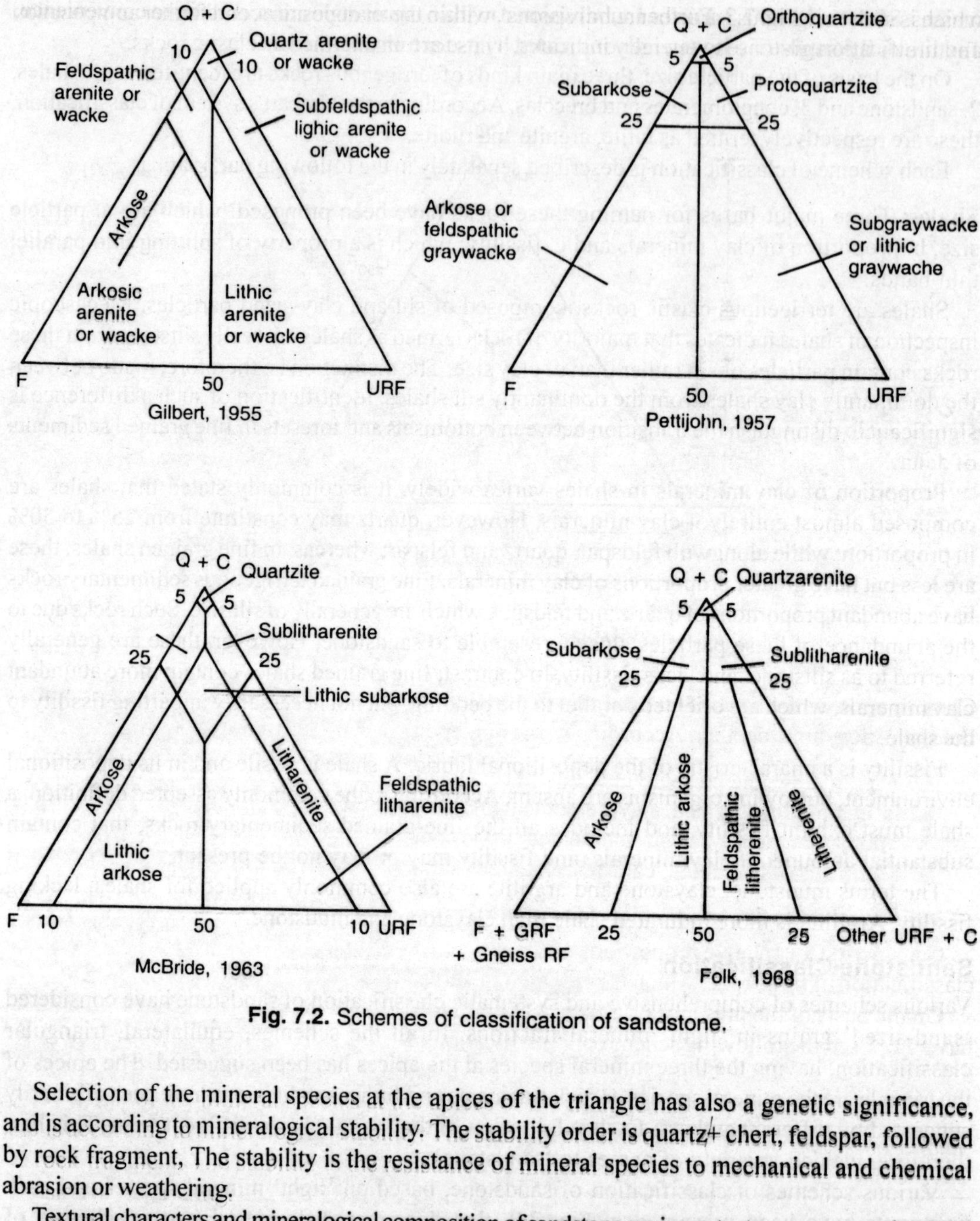

Fig. 7.2. Schemes of classification of sandstone.

Selection of the mineral species at the apices of the triangle has also a genetic significance, and is according to mineralogical stability. The stability order is quartz+ chert, feldspar, followed by rock fragment, The stability is the resistance of mineral species to mechanical and chemical abrasion or weathering.

Textural characters and mineralogical composition of sandstone are important aspects to understand its genesis and therefore are useful parameters in its classification.

Distinction in mineralogical composition is possible amongst quartz + chert, feldspar and rock or lithic fragment in microsection and also in handspecimen. Most of the schemes of classification, therefore, have used these three as apices of a triangle and have proposed triangular classification, which is shown in Fig. 7.3. Further subdivisions, within the triangle are according to convenience, Texture of the sandstone is generally indicated by its textural maturity.

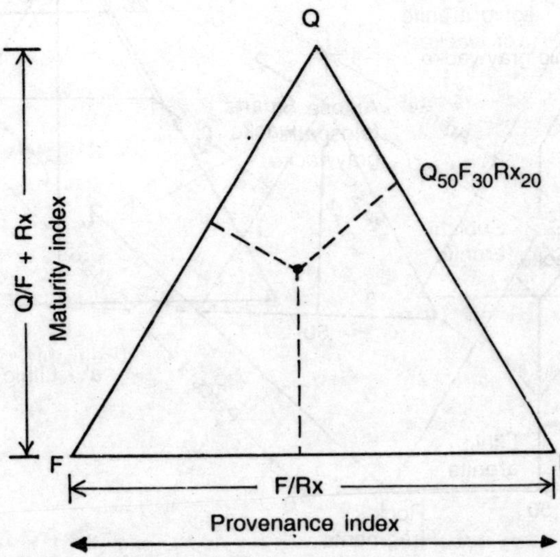

Fig. 7.3. Three components representing provenance index and maturity index for sandstone classification.

Classification on the basis of textural maturity indicates the rock name and the type of cement that holds clastic grains together. Accordingly, the nomenclature employed is as follows:

1. Supermature, medium-grained, quartz-cemented: quartz arenite;
2. Immature, fine grained, clay-cemented: arkose;
3. Submature, pebbly, calcite-cemented: lithic conglomerate;
4. Mature, coarse grained, calcite-cemented: subarkose.

Various schemes of classification of sandstone in common practice are presented in Fig. 7.4.

Considering the relative proportions of various constituents, Pettijohn (1957) has proposed the classification of sandstone, which is presented in Table 9.2.

On the basis of the presence or absence of argillaceous matter, Friedman and Sanders (1978) have proposed the following scheme of classification of sandstone.

Idealised end Member Rock Names

End member sand size particle	*<15% matrix*	*>15% matrix*
Quartz	Quartz sandstone Sandstone	Argillaceous Quartz
Feldspar	Feldspar Sandstone	Argillaceous Feldspar Sandstone
Rock Fragments	Rock fragment Sandstone	Argillaceous rock Fragment sandstone

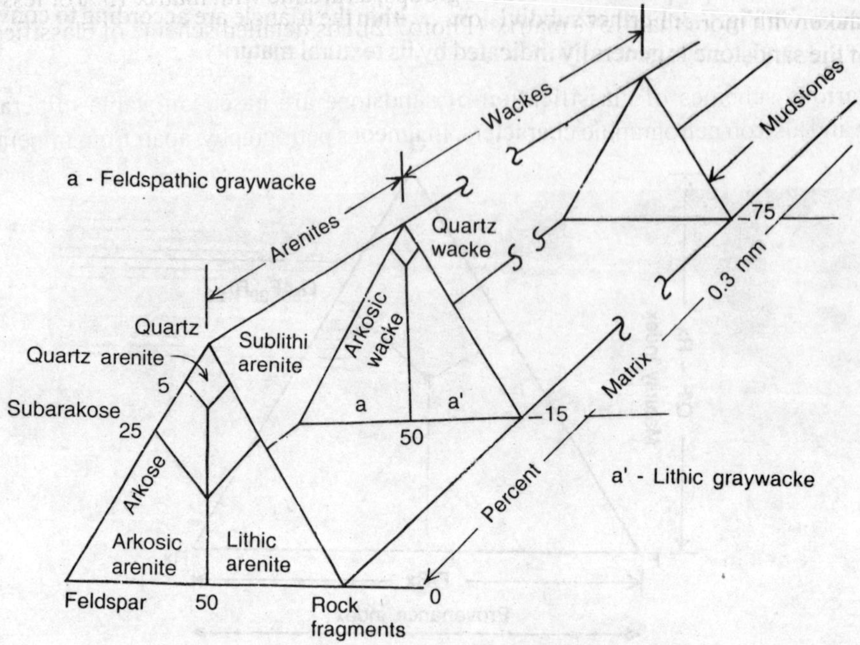

Fig. 7.4. Classification of terrigenous sandstone.

However, various schemes of classification of sandstone, outlined above, are primarily based on the earlier attempt made by Krynine, On the basis of a triangular diagram, he introduces the terms, quartzite, graywacke and arkose; as varieties of sandstone. However, the last two terms have led to some confusion.

According to Krynine, a graywacke is a sandstone composed of angular quartz (and chert) particles and abundant metamorphic rock fragments, with little or no cement and feldspar, and containing more than 12% to 17% micas and chlorite (either in the clay matrix or as metamorphic rock fragments). In this definition, Krynine assumed that mica and chlorite between the clastic grains were indeed a mechanically deposited matrix and thus represented poor sorting, However, this basic assumption is not necessarily correct. To avoid the confusion between these two terms, the term 'wacke' instead of 'graywacke' is used. According to Krynine, the term 'arkose' has been applied to sandstone with more than 30 percent feldspar.

Krynine had put an emphasis on diastrophism. The presence of particles and clay fraction reveal:

(*a*) provenance,
(*b*) availability of mud in depositional environment,
(*c*) energy level of depositing medium and
(*d*) diagenetic formation of clay mineral cement.

Therefore, emphasis on diastrophism by Krynine in classifying sandstone has no strong base, The Krynine's scheme of classification is, therefore, considered to be of historical significance.

Okada (1970) has suggested a classification of sandstone by way of triangular diagram by selecting quartz, feldspar and rock fragments including chert as apices of the triangle. On the basis of matrix present in sandstone, he proposed two broader groups, as arenite with matrix 15% or less (Photo 7.1) and Wacke with more than 15% matrix (Photo7.2). His detailed scheme of classification is presented in Fig. 7.5.

These various schemes of classification of sandstone are based on stable mineralogival assemblage and also on petrographic characters, In igneous petrography, apart from mineralogical

Photo 7.1. Sandstone showing arenitic framework.

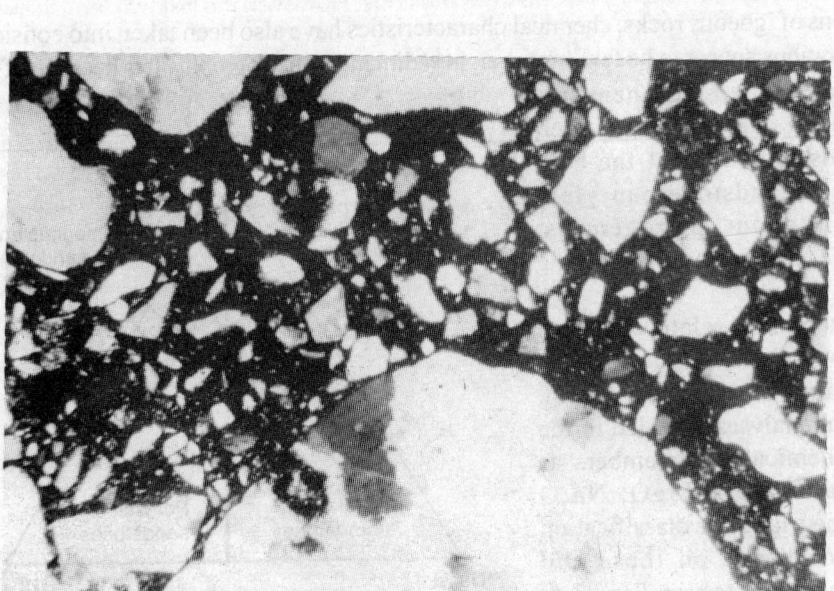

Photo 7.2. Sandstone showing wacke framework.

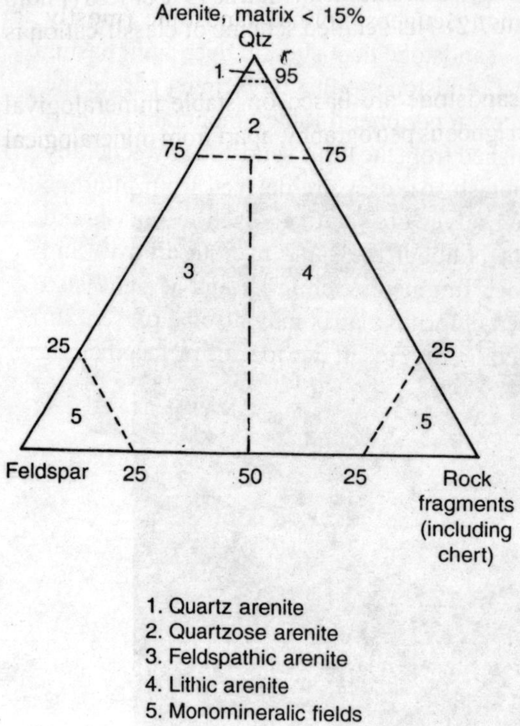

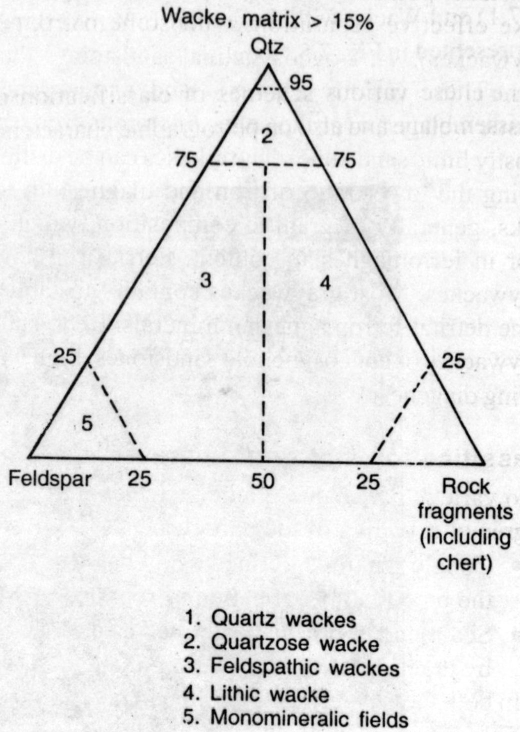

Fig. 7.5. Classification of sandstone (after Okada, 1971).

classifications of igneous rocks, chemical characteristics have also been taken into consideration. Such cnsiderations appear to be few, however, in sedimentary petrography. It is because of the post-depositional changes in chemistry involved during diagenesis, However, recent trends indicate that the bulk chemistry of sandstone can yield information about petrogenesis, including the changes involved during diagenesis.

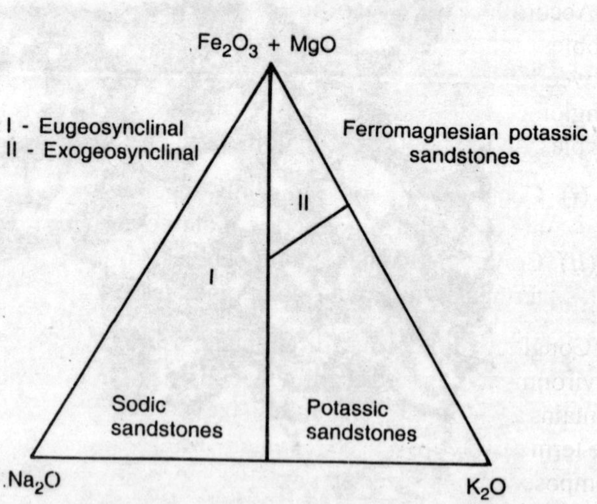

Fig. 7.6. Chemical classification of sandstone in relation to tectonic setting.

In an attempt to relate chemical composition of sandstones to the tectonic setting in which they form, mathematical analyses have led to the choice of chemical end members as total iron as Fe_2O_3+ MgO; Na_2O and K_2O. The triangular classification of sandstone based on these end members is presented in Fig. 7.6. Accordingly, three major kinds of sandstones have been identified as sodic sandstone, potassic sandstone and ferromagnesian potassic sandstone. This diagram hepls to make effective separation, with some overlap, among eugeosynclinal sandstone (mostly graywackes), taphrogeosynclinal sandstone—that is, sandstone deposited in deep fault basins on the craton (mostly arkoses)—and; exogeosynclinal sandstones—that is, forming part of the 'clastic wedges' over the edges of cratons from sources in peripheral folded mountain systems (mostly lithic sandstone). The arkoses can be distinguished from the lithic sandstone by generally having the lower ratio of iron and magnesium to potash. Arkoses are derived from plutoinc rocks, generally of granitic composition, which themselves are relatively rich in potash and poor in ferromagnesian content. Ferric iron content is about the same in both arkoses and graywackes. Most graywackes contain 'unstable' rock fragments rich in Fe and Mg and also some detrital ferromagnesian minerals. Recrystallised chloritic matrix may also be present. In graywackes or eugeosynclinal sandstones, high Na/K ratio may result due to soda metasomatism during diagenesis.

Classification of Conglomerates and Breccias

Two varieties of coarse-grained terrigenous clastic rocks are recognised that are based on the degree of rounding of the particles.

- Conglomerate is defined as a coarse grained terrigenous sedimenatary rock, formed by the process of lithification of rounded gravel.
- Sedimentary breccia is defined as a coarse grained terrigenous sedimentary rock, formed by the process of lithification of angular gravel.

In both the types, more than 30 percent of the large particles exceed 2 mm (–1ϕ) diameter. The particles may be pebbles, cobbles, boulders or mixtures of the particles of these various sizes. Conglomerates and sedimentary breccias are named and classified on the basis of:

(*i*) the proportion of gravel-sized particles;
(*ii*) the kind of matrix and
(*iii*) the kinds of gravel-sized particles.

According to the proportion of gravel-sized particles and the kind of matrix; sample with 80% pebbles, cobbles and/or boulders is a conglomerate proper; while the one having such particles between 30% and 80% is a sandy or arenaceous conglomerate or a shaley or argillaceous conglomerate. The matrix may be calcareous or sideritic. On the basis of the variety of pebbles, cobbles, and/or boulders, conglomerate can be classified into two types as,

(*i*) Conglomerate consisting only of pebbles, cobbles, and/or boulders of a single kind of rock are termed as oligomictic conglomerate; and
(*ii*) Conglomerate containing pebbles, cobbles, and/or boulders of many kinds of rocks are termed as polymictic conglomerate.

Conglomerates form in different depositional environments, Accordingly, the one from glacial environment is termed as tillite. It consists of till, which is a diamicton, deposited by glacier. Till contains a wide variety of particle sizes. To offset the term tillite, as it denotes genetic significance, the term diamictite is used for a non-sorted, non-calcareous, terrigenous sedimentary rock that is composed of sand sized and/or large sized particles, distributed in fine-grained matrix.

Similarly, conglomerates from alluvial fans are termed as fanglomerates.

7.4 CLASSIFYING INTRABASINAL OR AUTOCHTHONOUS ROCKS

It has been estimated that nearly 70% of the Earth's surface is underlain by strata composed of sedimentary rocks. The aggregate thickness of sedimentary strata locally may range upto 15 km. and more. Of the entire sedimentary strata, carbonate rocks, both limestone and dolostone, constitute nearly 20% of the total. Schemes of classification of these rocks are presented in the following paragraphs.

Classifying and Naming Carbonate Rocks

Limestones by definition are carbonate rocks which contain more than 50% carbonate minerals. There are two broader compositional divisions, as, limestones and dolostones. Dolostones are carbonate rocks that consist of the mineral-dolomite

Three different schemes of classification are descriptive and have genetic implications. Each scheme of classification is based on three different constituents, namely: (*i*) sand-sized particles, (*ii*) micrite and (*iii*) sparry cement.

Micrite is lithified, mechanically deposited lime mud, which occurs either as a matrix in sand-sized particles in a limestone or as the only particles in a fine-grained limestone. Particle size in micrite is 4μ but more commonly lies between 1 μ and 3μ. Sparry cement is chemically precipitated, pore-filling cement and occupies interstices amongst the sand-sized particles. Folk (1959) has designated sand sized particles as allochemical constituents or simply as allochems; while micrite and sarry cement have been designated as orthochemical constituents, or simply as orthochems.

The three different schemes of classification have been suggested by: (*i*) Leighton-Pendexter (*ii*) R. J. Dunham and (*iii*) R. L. Folk. These are described in the following paragraphs.

Classification of the Limestone by Leighton-Pendexter

While classifying limestones, it is recognised that size frequency distribution of the particles is generally bimodal, and limestone consists primarily of (*i*) coarse particles, generally of sand size or larger; and (*ii*) fine particles of silt and clay size, Leighton and Pendexter classified limestones and named them according to the relative proportions of coarse particles and micrite This group of carbonate has been separated from rocks produced by organic framework builders, viz.; reefs and deposits, such as; caliche, travertine and tufa, The coarse particles include detrital and skeletal particles, pellets, lumps and coated particles.

The term detrital includes intraclasts and also fragments derived from older limestones. Skeletal particles are whole or broken tests of organisms; pellets may be faecal or pelloids or earlier intraclasts; whereas lumps include composite grains, like grapestones and oncolites. Coated particles are oolites and pisolites and particles encrusted with forms. The classification suggested by Leighton and Pendexter is presented in Table 7.2.

If there are more than one kind of particles, name of each is included in clssification. The order in naming the rock is based on relative abundance of each contituents, the first being the abundant kind of particle, followed by next in abundance, while name of the least abundant particle is mentioned last. One of the advantages of this classification is the use of simple terminology.

Classification of the Limestone by Dunham

For the purpose of classification, Dunham has distinguished two kinds of limestones, on the basis of:

(*i*) the original components bound together, and

TABLE 7.2: Classification of Limestone (after Leighton and Pendexter)

		Grain Type						
Grain/ micrite ratio	*% grains*	*Detrital grains*	*Skeletal grains*	*Pellets*	*Lumps*	*Coated grains*	*Organic frame builders*	*No organic frame builders*
9 : 1	90%	Detrital L.S.	Skeletal L.S.	Pellet L.S.	Lump L.S.	Oolitic L.S. Pisolitic L.S. Algal. encru. L.S.	Coralline L.S. Algal L.S.	
1 : 1	50 %	Detrital micritic L.S.	Skeletal micritic L.S.	Pellet micritic L.S.	Lump micritic L.S.	Oolitic (pisolitic) micritic L.S.	Coralline micritic L.S. Algal micritic L.S.	Caliche Travertine Tufa
1 : 9	10 %	Micritic detrital L.S.	Micritic skeletal L.S.	Micritic pellet L.S.	Micritic lump L.S.	Micritic oolitic (pisolitic) L.S.	Micritic coralline L.S. micritic algal L.S.	
				Micritic Limestone				

(*ii*) in which the original components are not bound together.

The first group of limestones exhibits evidence of binding during deposition, i.e boundstone. This group includes reefs, stromatolites and travertine.

The second group of limestones includes coarse particles and micrite.

Dunham considered the abundance of coarse particles and micrite as the major criterion in classifying and naming limestone. He used the term lime mud instead of micrite. Relationship between sand-sized particles and lime-mud helps to distinguish an original sediment, deposited in clean water from that of sediment, deposited in agitated water, In clean water, lime mud is able to settle on the bottom and remains there, and rocks that form original lime mud remain contrasted with those originally devoid of lime mud; irrespective of number and size of sand-sized particles.

According to Dunham, there are two categories of limestone, which are not bound together during deposition:

(*i*) limestones in the depositional environment composed of sand-sized and larger particles, which supported one another, i.e. particle supported, and
(*ii*) limestones in the depositional environment consist of lime mud in which particles are enclosed or 'floated' and did not support one another, i.e. mud supported.

Particle-supported limestones fall into two categories: (a) particle-supported limestones entirely lacking in lime mud, termed as grainstones; and (b) limestones containing lime mud, but those particle supported, termed as packstones. Such mud supported limestones are further subdivided into two categories, as, (a) mud-supported limestone containing more than 10% sand-size or larger particles; termed as mud stones; and (b) mud-supported limestones containing more than 10% sand size or larger particles, known as wackestone. Limestone classification suggested by Dunham is presented in Table 7.3.

TABLE 7.3: Classification of Limestone (after Dunham)

Depositional Texture				
Original components not bound together during deposition				Original components bound together during deposition
Contains mud			Lacks mud and is grain supported	
Mud supported		Grain supported		
<10%	>10%			
Grains	Grains	Packstone	Grainstone	Boundstone
Mud stone	Wackestone			

Classification of Limestone by Folk

As per Leighton and Pendexter, Folk had also considered that limestones are bimodal in size distribution; i.e. they are composed of $CaCO_3$ grains, usually of coarse silt or sand size, and carbonate mud, generally of clay size.

Folk had identified that majority of carbonate rocks consists of three components as:

(*i*) discrete carbonate aggregates, termed as allochems, including particles, such as; intraclasts, ooliths, skeletal grains and pellets,

(*ii*) micro crystalline calcite, i.e. micrite or carbonate mud and

(*iii*) sparry calcite, a chemically precipitated porefilling cement.

The last two are termed as orthochems. On the basis of combinations of sand-size or larger particles with micrite or cement, eight different kinds of limestones have been suggestd. In addition, micrite may lack sand-size particles, giving rise to ninth kind of limestone and tenth kind of rock is identified as reef rock.

Accordingly, on the basis of three framework constituents and their relative proportions, Folk had suggested the classification of carbonate rocks, as presented in Table 7.4.

TABLE 7.4: Classification of limestone (after Folk)

Allochemical Rock			*Orthochemical rock*	*Reef rock*
Sparry	Calcite Cement	Micrite		
Intraclasts	Intrasparite	Intramicrite	Micrite	Biolithite
Ooids	Oosparite	Oomicrite		
Fossils	Biosparite	Biomicrite		
Pellets	Pelsparite	Pelmicrite		

These are the various schemes of classification of carbonate rocks.

Classification of Carbonaceous Rocks

Carbonaceous rocks, namely the coal rock, are classified in two ways; (*i*) by rank, i.e. content of carbon and calorific value, and ii. by petrographic characteristics. (*i*) Rank of coal is a measure of diagenetic and metamorphic grade that reflects the pressure, shear and temperature to which the coal was subjected to. In order of increasing rank, the classes of coal have been identified as, (*i*) lignite, (*ii*) sub-bituminous coal, (*iii*) bituminous coal and (*iv*) anthracite.

The progression of carbonaceous material from lignite or its earlier stage as peat to anthracite is known as the process of coalification. It is accomplished first by biochemical and later by geochemical processes. In the conversion of plant material to coal, there is an increase of carbon and calorific value, with a corresponding decrease in moisture, volatile matter and oxygen.

On the basis of petrographic characters, four different kinds of coal are identified, as; (*i*) vitrain or anthraxylon: It is a bright, glassy, vitreous or glassy-looking coal. It has an earthy appearance, but lacks in lustre; (*ii*) Durain: It is dull coal, lacking lustre and having an earthy appearance; (*iii*) Clarain: It has a smooth surface. It is glossy and shining and exhibits silky lustre. (*iv*) Fusain: ("mineral charcoal" or "mother of coal"). It is carbonized wood that resembles charcoal. When touched, it soils the hand.

These four kinds of coal can be easily studied by examination of polished surfaces, under the microscope using reflected light. However, these four kinds rarely occur as independent type, but are more commonly interlaminated.

Another group of Intrabasinal rocks is 'authigenic' rocks. These are chemically and mineralogically quite diverse. Of these,important kinds are i. chert, ii. evaporite iii. phosphate rocks and iv. iron and manganese rocks.

These types of authigenic rocks have been discussed later at length in the respective chapters.

7.5 CLASSIFYING PYROCLASTIC ROCKS

Pyroclastic rocks are lithified tephra, that originate as explosive, igneous material out of the volcanic eruption, but are deposited as sediments. Volcanoes and their products, both on land and in ocean, are the major source of particles that form pyroclastic rocks. The areas of tectonic activity and island arcs are favourable environment for their formation.

These rocks are classified on the basis of (*i*) size, (*ii*) kinds of particles composing them and (*iii*) origin.

On the basis of size, pyroclastic particles are categorised into three types, as; (*a*) particles larger than 64 mm, (*b*) particles between 64 mm. and 2 mm., and (*c*) particles smaller than 2 mm. Particles larger than 64 mm. in diameter are further subdivided into two groups, on the basis of shape. These are bombs, particles of 64 mm. size, commonly exhibiting twisted forms and concentric structures because these are ejected as liquid lava and consolidated during the flight and blocks, particles having 64 mm. diameter, ejected from volcano as solids and consist of angular fragments of volcanic rocks.

Particles in the intermediate size range are termed as lapilli, which are the pyroclastic particles with 2 to 64 mm. diameter. Particles in the fine size range are termed as ash, and which are pyroclastic particles having diameter less than 2 mm.

Pyroclastic rocks, consisting mostly of bombs, are termed as agglomerates; while those consisting mostly of blocks are termed as volcanic breccia.

Pyroclastic rocks consisting mostly of lapilli are named lapilli tuffs; and those consisting of ash are termed as tuffs, in the strict sense.

To classify pyroclastic rocks on the basis of composition is rather difficult, as majority of them are fine grained and glassy. According to the composition, there are three end members, as: (*i*) rock fragments, (*ii*) crystals, and (*iii*) glass.

On the basis of composition, pyroclastic rocks are classified by applying a suitable adjective and naming it as a crystal tuff, rock fragment tuff or glassy tuff.

A variety of tuff that carries a special name is ignimbrite or welded tuff. It is a nonsorted pyroclastic rock, deposited while the ejected particles were still in a plastic condition.

8 CHAPTER

Sedimentary Facies and Products

8.1 INTRODUCTION

Studies indicate that sedimentary rocks cover nearly 66% of the continental areas and practically most of the ocean floor. Of the sedimentary rocks, sandstones, mud rocks and carbonate rocks constitute nearly 95%. Analysis of stratigraphic record of the preserved sedimentary sequences indicate that majority of them are marine, though exact proportion of marine and non-marine rocks has not been ascertained.

Earlier to the concept of Plate Tectonics, thickest sequences of marine sediments were considered to have been deposited. in a generally linear, areally extensive, structural depressions, termed as 'Geosyncline'. These have been located at the edges of continents. Such a geosyncline has further been divided into two parallel segments, termed as miogeosyncline and eugeosyncline. Miogeosyncline is adjacent to continent and generally characterised by shallow marine sediments, lacking in volcanic rocks or volcanic detritus; whereas eugeosyncline is characterised rather by deep water areas, and is situated away from the continents or continental margins. It has thick accumulations of sediments of real marine character, which, in turn, are associated or interstratified with volcanic rocks or volcanic detritus. Idealised examples of such thick accumulations of sedimentary sequences deposited in geosynclines are Appalachian mountains from North America and the Alpine-Himalayan folded mountain chain. Geosynclinal theory has gradually been modified in the light of Plate Tectonics.

Evolution of sedimentary basin is related to the formation and consumption of lithosphere, types of plate interactions as convergent, divergent or transform fault and its location with respect to plate boundaries. However, sequence of evolution of sedimentary basin in space and time varies within wide limits. Nature of basin evolution is based on its location, and regarded as continuous interaction between tectonics and sedimentation. The sediment depositional sites, i.e. sedimentary basins that are identified are as follows:

(*a*) Oceanic basins underlain by oceanic lithosphere,

(*b*) Shifted continental margins along the contact between oceanic and continental lithosphere (Atlantic-type margin), (*c*) Arc-trench systems, where oceanic lithosphere is consumed beneath island arcs or continental margins (Island arc or Andean type margin),

(*d*) Suture belts, where continental blocks are juxtaposed by crustal collision,

(*e*) Intra-continental basins in the interior of continental blocks.

There are the five broader sites, identified on the earth's surface (Fig. 8.1), where sediments accumulate and are transformed later into sedimentary rocks. These are identified as 'Depositional Environments'. A depositional environment can therefore be defined as a natural geographic entity, in which sediments accumulate. Both the processes of sedimentation and deposition are controlled by tectonic processes, whereas each depositional environment is characterised by physical, chemical and also organic processes. A final sedimentary product, therefore, is a complex of interplay between physical, chemical and tectonic rocesses. This results in a close relation betweeen sedimentary product and environment of deposition.

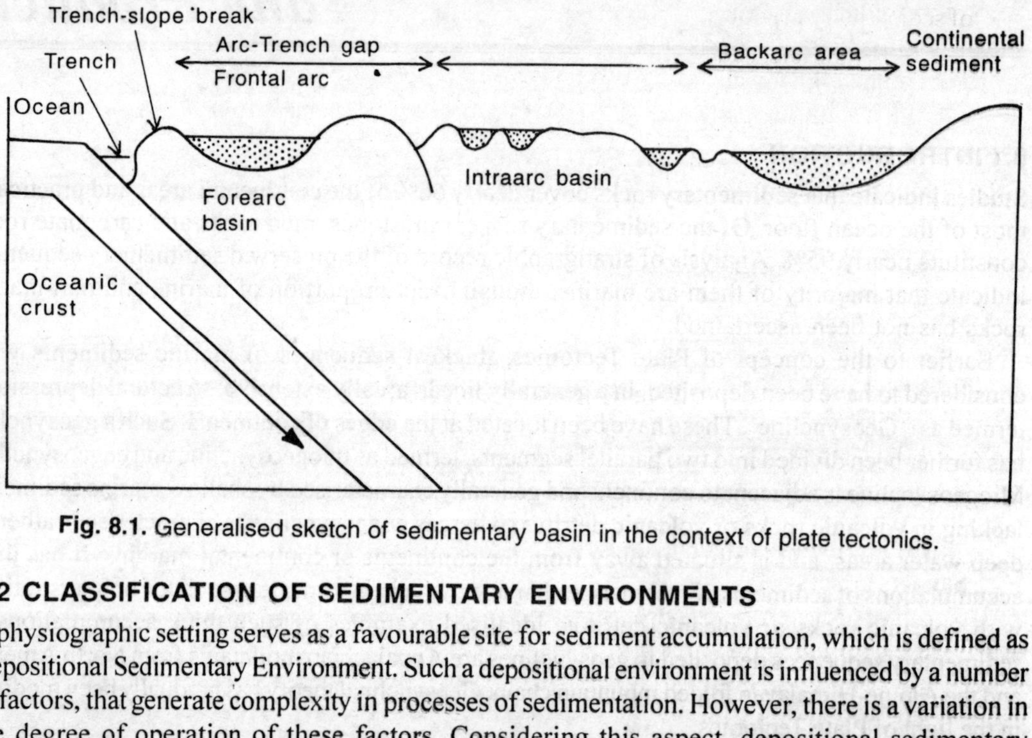

Fig. 8.1. Generalised sketch of sedimentary basin in the context of plate tectonics.

8.2 CLASSIFICATION OF SEDIMENTARY ENVIRONMENTS

A physiographic setting serves as a favourable site for sediment accumulation, which is defined as Depositional Sedimentary Environment. Such a depositional environment is influenced by a number of factors, that generate complexity in processes of sedimentation. However, there is a variation in the degree of operation of these factors. Considering this aspect, depositional sedimentary environments are classified in the light of its location with respect to plate tectonic setting, as described in the earlier section.

Sedimentary environments have also been classified on the bases of physical setting, tectonics and climate, as controlling factors. These environments are described in the form of different facies models. Such a scheme of classification appears to be a simpler one.

Broadly, two physiographic settings of depositional environment are identified, viz; continental and oceanic areas on the earth's surface. Accordingly, continental or non-marine and oceanic or marine are the areas of sediment deposition. Environments which are not truly continental or also truly oceanic or marine are termed as transitional environments. In such a transitional environment, there is an interplay of both, i.e. continental and marine processes. Classification of depositional environments is made on similar lines, which are described as facies model. It has been defined as a natural geographic entity in which sediments accumulate, which are broadly grouped into (i) marine, (ii) non-marine, and (iii) transitional environments (Fig. 8.2).

8.3 MARINE ENVIRONMENTS

Information and knowledge about marine environments have been obtained by systematic study of modern oceans.The marine environment is distinguished by chemical characteristics of its waters, by enormous variations in its depths, by distinctive temperature structure that bears a close relation to global climatic zones, by various physical and dynamic processes and also by varying groups of organisms. Such characteristics of modern oceanic environments and a great range of variations in them are also affected by adjacent continental areas. Therefore, there exists a variety of sedimentary products.

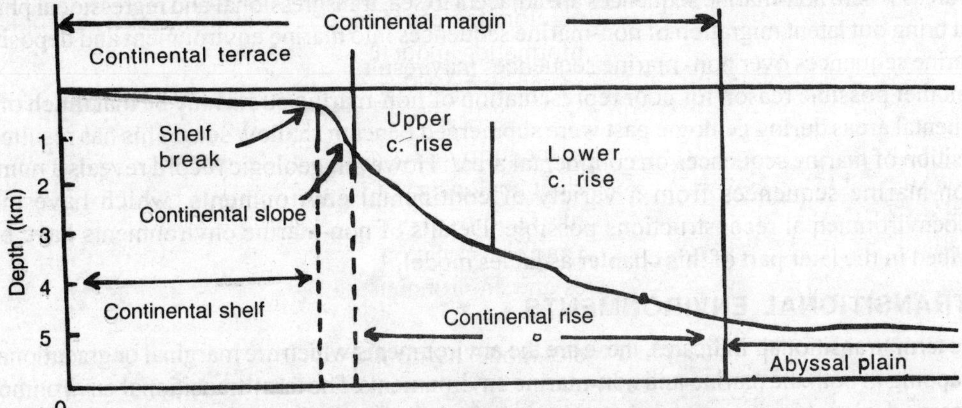

Fig. 8.2. Broader marine environments.

The marine environment has further been broadly divided into two great realms (Friedman and Sanders, 1978), as:

(*i*) the benthonic realm : It is a collective designation for all the deeper areas of ocean floor; and

(*ii*) the pelagic realm, that includes all the ocean water, which lies seaward of low-tide level.

The pelagic realm, in turn, consists of two major provinces, as:

(*a*) neritic province, i.e. the coastal-ocean water that overlies continental shelves; and

(*b*) oceanic province, i.e. blue-water region, overlying deep-sea basins.

Studies have been made to understand nature of various processes that operate in modern marine environments, which in turn help to analyse the geologic record on sound basis.

Marine environments and various sedimentary products associated with them have been dealt with in the later part of this chapter, where facies models have been described at length.

8.4 NON-MARINE ENVIRONMENTS

These are the environments, where the sedimentary products are deposited in arid and semi-arid regions and include a variety of environments. These environments are of great diversity and of ecological variability. The products of these environments have been studied in depth and therefore these non-marine environments are best known to earth scientist; even though these are least represented in the geologic record.

Low proportion of non-marine strata in the geologic record is possibly because non-marine environments are situated above the base level of erosion and their products therefore get easily eroded and may leave no trace of their occurrence. When the basement beneath the non-marine environment subsides greatly, it is a 'temporary' site of deposition. Majority of the products of non-marine environments are predominant sources of sediments, which are later deposited in marine environment. If the subsidence of non-marine sequence takes place more rapidly than it gets eroded, there is a possibility of their preservation in the stratigraphic record. More commonly, non-marine sequences of sedimentary rocks have been preserved in the areas of rifts or grabens and of great downwarps. However, for the supply of vast amounts of sediments to fill such areas, there was a source area that provided required sediments.

In areas where non-marine sequences are adjacent to sea, transgressional and regressional phases of sea bring out latent migration of non-marine sequences into marine environment and deposition of marine sequences over non- marine sequences may result.

Another possible reason for poor representation of non-marine strata may be that much of the continental areas during geologic past were submerged beneath shallow seas. This has resulted in deposition of marine sequences on continental areas. However, geologic record reveals a number of non-marine sequences from a variety of continental environments, which have made palaeoenvironmental reconstructions possible. Details of non-marine environments have been described in the later part of this chapter as facies model.

8.5 TRANSITIONAL ENVIRONMENTS

As the term 'transitional' indicates, these are the environments which are marginal or gradational or overlapping in both the marine and non-marine environments.The term 'transitional environment', therefore, designates numerous places in marginal parts of the sea, where accumulation of terrigenous sediments can take place. In coastal areas, there are distinctive water current patterns, which control distribution of sediments. These sediments and sediment distribution pattern are distinctive by themselves, as some of the sediments are land-derived, while some are oceanic. In transitional environment, therefore, marine and non-marine processes interact. Marine processes are mainly controlled by salinity and circulation of sea water and by waves, tides and currents, while major non-marine processes are controlled by fresh water flows entering into sea through fluvial system. Resulting sedimentary products, therefore, bear imprints of both i.e. marine and nonmarine environments. Physiographic setting of a sedimentary product deciphers a relatively significant imprint of either marine or non-marine processes.

Details of the sequences of transitional environment have been discussed at length in the later part of this chapter, as sedimentary facies models.

8.6 FACIES CONCEPT AND FACIES MODELS

From the preceding paragraphs, it can be stated that a sedimentary product is characterised by specific lithologic and biologic properties that are imparted to it by various parameters of the environment of deposition. Such a depositional environment is defined at any one time by geographical boundaries, which determine lateral or horizontal dimensions of the sedimentary deposits. Therefore, there exists the close relation between the sedimentary deposit and environment of deposition. This sedimentary deposit is termed as; 'facies'. Definition of 'facies', as suggested by Moore (1949), has been reproduced by Blatt, *et al.* (1980). It reads as, "Sedimentary facies is defined as any areally restricted part of a designated stratigraphic unit, which exhibits characters significantly different from those of other parts of the unit". It is observed that geographical boundaries of the depositional environment change or migrate from time to time. The characteristic sedimentary deposits or facies, therefore, spread laterally with time.

Sedimentary facies is further divisible into units or subfacies, which differ from one another in either lithological or biological characters.

Sedimentary facies are restricted in extent both stratigraphically and geographically, eventhough the same facies is encountered at different levels within the same stratigraphic unit. It is generally understood that these units are ultimately given an environmental interpretation.

It is observed that facies in one stratigraphic unit may bear close resemblance with other units that are of different age and also occur in different parts of the world. This leads to suggest that the term 'facies' in more general sense is an association of sedimentary rocks that have some aspects common in appearance. Accordingly, it can be described as 'a white sandstone facies'. It is therefore termed as Descriptive Sedimentary Facies. On the basis of the environment of deposition, a facies can be termed as 'alluvial fan facies', which is designated as Interpretative Sedimentary Facies. However, in a number of cases, environmental interpretation may not prove to be correct. It is, therefore, advisable that facies be fully described and a purely descriptive designation be given, before assigning any environmental interpretation.

As stated, the term facies is used in different ways, and accordingly suites of sedimentary rocks are assigned to different facies. However, there are two suites of terrigenous sedimentary rocks with striking contrast. Swiss geologists have termed one of the suites, as; 'flysch' and another as 'molasse'.

According to definition, 'flysch' is a thick succession of marine sedimentary strata, which consist of repetitively interbedded alternating and laterally persistent sands and/or coarse sediments and shales found in the interior of a folded moutain chain.

There is a peculiar variety of suite of rocks that is associated with 'ordinary' flysch and it is termed as 'wildflysch'. It consists of blocks of igneous-, sedimentary-, and metamorphic rocks of various dimensions, embedded in a very fine matrix of typically dark-coloured marine shales or siltstone or mudstone. In the present day terminology, it is a kind of diamictite.

In case of flysch, its environment of deposition has always been debatable. Presence of coarse layers of sediments, ripple marks and plant debris points towards shallow-water environments of deposition, whereas thick sequences of sediments of tens of thousands of meter in Flysch contradict the shallow water origin. Environment of deposition of flysch has, therefore, remained an enigma. However, earlier geologists identified its relation with 'tectonics' and was considered as, pre-orogenic phase of sedimentation, i.e. before the geosynclinal phase culminated into the folded mountain chain. To account for its shallow-water characteristics, simultaneous process of sedimentation and subsidence has been suggested under the Geosynclinal theory. Sedimentation had resulted due to turbidites, in turn, to form synorogenic flysch sequences.

Another suite of rocks with striking contrast is termed as 'Molasse'. It can be referred to as a tectono-stratigraphic unit. It is a wedge shaped body of sediments of extra-basinal origin. It is a patterned succession of non-marine and shallow marine strata. Constituents or particles of the sequence are derived due to erosion of the pre-existing rocks, including flysch, which build-up up-lifted folded mountain chain. Sediments of the molasse are of near shore origin and are considered to be autocyclic and are deposited in the foreland basin, i.e. shallow sea in front of the folded mountain chain. Foreland basin, as a shallow sea, is filled by fine sediments of fluvial and deltaic nature. Considering its relation with tectonics, molasse is regarded as a 'post-orogenic' deposit. An alternative term 'tectonic fan delta complex' is also applicable to molasse.

As stated in the preceding paragraphs that the shift in geographical boundaries of the depositional environment decipher horizontal or lateral extent of the sedimentary facies. Even the size of sedimentation, i.e. the position of depositional environment may change or shift through time. This results into the shift of sedimentary products, which in turn may overlie those of adjacent environments. If analysed in vertical section, sedimentary facies of both the environments appear interbedded, and while analysed in areal, lateral extent, facies may interfinger with each other. Thus, fluvial facies may interfinger with coastal-dune facies, which, in turn, may interfinger with beach facies and further into shallow marine facies or vice versa.

It has been noted that continental margins were rapidly submerged during the last 13,000 years due to deglaciation of the polar ice, termed as 'flandrian transgression'. This has given rise to interfingering of beach and non-marine facies in the coastal zone.

Thus, depositional environments and therefore facies are subject to controls due to geological setting, tectonics and climate. Therefore, information about the internal characteristics and spatial relations of facies are obtained and compared with that derived from modern environmental analogs. Thus, present is used to interpret past.

To carry out such comparative studies, the term *facies analysis* is used and generalisations based on various similar modern and ancient analogs, are referred to as facies model. A facies model is therefore constructed, which has the features of general rather than of local significance.

Facies analysis includes the description and classification of any sedimentary deposit It is then followed by interpretation of its processes and environments of depossition. This leads to the construction of facies model.

According to Walker (1976), a good facies model has four following functions;

(*a*) It can serve as a norm with which any real example can be compared;
(*b*) It can act as a framework and guide for future observations;
(*c*) It can act as a predictor in new geological situations and
(*d*) It can act as a basis for interpretation of the facies in terms of physical or chemical mechanisms.

Classification of facies model is both simpler and also more complicated than classification of environments. It is simple because each model incorporates related environments, while it is more complicated because the model is to be based not only on environments but also on other major controls, such as; climate, tectonics and physical setting. On the basis of these factors, Blatt, *et al.* (1980) have suggested classification of facies models. It is reproduced, in brief, in Table 8.1.

In classifying sedimentary environments of deposition or facies models, continental environments have been termed as non-marine, continental margin as transitional environment and oceanic as marine environment. These are described, in brief, in the following paragraphs.

A. Non-marine—Continental Facies Models

1. Alluvial Fans

An alluvial fan is a fan-shaped body that consists of coarse, detrital, poorly sorted sediments. It is built by a stream, originating at the base of a mountain front. At this front, steeper slope passes abruptly into a more gentle slope.

TABLE 8.1: Classifications of Facies Models

A. Predominantly continental
1. Alluvial Fans
2. Fluvial Plains
3. Sandy Deserts
4. Lakes
5. Coasts (non deltaic)
6. Deltas
7. Glacial and Glacial Marine
8. Terrigenous Shelf or Shallow Sea
9. Carbonate Shelf or Platform

B. Continental margin
10. Slope, Rise, and Basin Plain (mainly stable continental margins)
11. Subducting Plate Margins.

C. Oceanic
12. Deep (oceanic) Ridges and Basins

Alluvial fans are encountered on many different scales and occur in humid as well as arid and semi-arid climatic regions. Alluvial plains of the Ganges river at the Himalayan mountain front is an example from humid tropical areas. Generally, individual fans from semi-arid regions have areas of 1 to 100 km^2, but the Himalayan fans have much larger areas and may be as much as 150 km. across streams, originating from a mountain front to give rise to fans. Such streams generally coalesce to form a surface that slopes away from the mountain front.

At the origin of the mountainous stream, i.e., at the fan-head segment, slopes are steepest and flow is normally confined to a single stream or channel, that may get entrenched due to stream action. Entrenchment may be shallow or deep. Shallow entrenchment may be temporary and back-filling or clogging of the channel may occur due to debris flow deposits. Deep entrenchment may lead to bypassing of the fan head and deposition of sediments results into shallowing of the channel particularly in the mid fan or lower fan regions.

Deposition of sediments on the fan surface generally takes place by several processes. Debris flow and upper regime stream flow take place in arid or semiarid climatic regions. Thin and relatively unconfined upper regime flows are sheet-floods and result due to local intense rains. Debris flows are generally confined to fan head and mid fan, whereas, more mobile stream floods may reach lower fan. In glacial and humid climate, however, debris flows are not generally formed and deposition is usually due to stream floods and braided stream flows. Predominance of stream floods results into the sediments, which are characteristics of coarse braided streams. Alluvial fans from semiarid regions are characterised by steep slopes.Initially, eposition of sediments is restricted to narrow and shallow channels of the fans, which may shift gradually, as the shallow channels get filled by sediments.

Apex of the fan is characterised by coarse, poorly sorted debris flow deposits. Maximum clast size gradually reduces away from the apex, in the mid fan. Deposition of gravels in the mid fan also gives rise to well-developed imbrication, having orientation of pebble long axes across or downslope, i.e. either transverse or parallel to the current direction. Fig. 8.3 shows schematic diagram of the facies model.

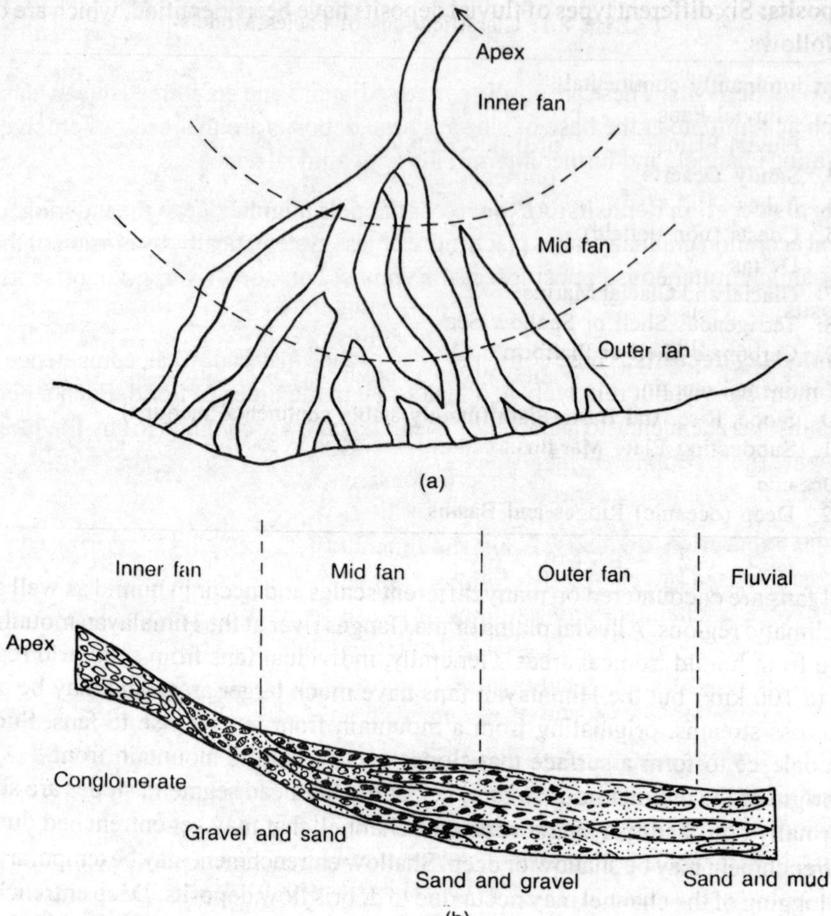

Fig. 8.3. Alluvial fan model. (a) Plan view; (b) Vertical section.

2. Fluvial Plains

Sediment deposits of this facies model are of broad aggradational flood plains. Two distinct types of this model have been identified, which correspond according to meandering pattern and braided channel pattern.

Generally, flood plains of different rivers or of different streams merge with each other to give rise to broad alluvial plains. Majority of them are located on slowly subsiding continental margins and therefore, are bordered on one side by sea. Major rivers may develop deltas, where upper parts of the deltaic plains merge with the lower, meandering part of the river system. Alluvial plains are characterised by more than one river system, which flow more or less parallel to each other on a gentle, well-developed sloping surface. This surface is characterised by intertongued fluvial, deltaic and coastal marine sediments. The Ganges river is an example of the configuration of fluvial and deltaic systems. Its sediments accumulate in a subsiding trough i.e. foredeep on the craton side of a mountain arc, while flow of the major river is parallel to the mountain front. It is fed by tributories that develop alluvial cones, extending from the mountains.

Fluvial Deposits: Six different types of fluvial deposits have been identified, which are described, in brief, as follows:

1. **Colluvial deposits:** These are poorly sorted sediments and are formed due to mass wasting, which accumulate at the base of slope. These deposits are the result of erosive action of migrating channels and further intermixing with alluvial deposits.

2. **Lateral accretion deposits:** As the river channels migrate due to meandering, process of lateral accretion gradually takes place more or less systematically, by erosion of the concave bank and simultaneous deposition at the convex bank of a river, giving rise to point bar deposits.

3. **Channel lag deposits:** These consist of coarsest bed material. Weak competence of current shift them slowly, but relatiyely at a faster rate at the times of flood. Such a bed material generally has a tendency to lag behind. These deposits are common to any fluvial system but rather are more common in meandering river system. Such lag deposits may include blocks of overbank sediments and waterlogged plant debris.

4. **Vertical accretion deposits:** These sediment deposits form due to overbank flooding and are characterised by very fine sand and mud. These are carried and deposited in suspension, either close to channel margin (also where coarse fraction is deposited to form natural levee) or in relatively low lying areas on plains, such as, shallow lakes or backswamps. Due to these accretion deposits, aggradation of flood plains takes place. Generally meandering system is favourable for accretion rather than a braided system.

5. **Channel fill deposits:** These are formed if the channels are abandoned by certain development in a fluvial system. Such a development may result either by progressive reduction of discharge or by a sudden shift to a lower part of the flood plain or by meander cut off. In case where braided channels are aggrading, channels may get clogged and back-filled with coarse sediments. In case, where shift or meander cut-off has taken place, channel fill deposits include fine overbank mud and organic detritus.

6. **Floodplain splay deposits:** Development of these deposits is due to partial breaching of natural levee at the time of floods, which comprise bed material and wash load that gets carried out of the channel and is spreaded over the adjacent floodplains . The 'breaks' in natural levee are progressively filled by deposition of bed material. Splay deposits, therefore, typically exhibit thin fining upward sequences, intercalated in overbank, i.e. lake or swamp deposits.

Fig. 8.4 presents different types of fluvial deposits.

Channel Patterns: On the basis of the nature of fluvial channels, two distinct types (Fig. 8.5) have been identified:

(*i*) straight or meandering channel,in which the entire flow is normally restricted to a single channel, and

(*ii*) braided or anastomising channel; in which flow is divided into a network of branching and reuniting channels.

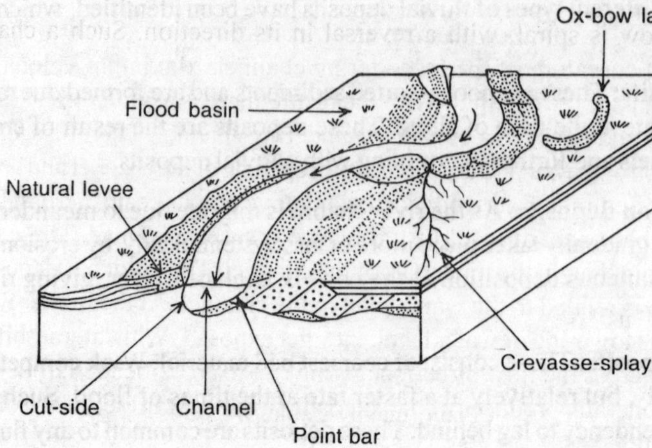

Fig. 8.4. Various types of fluvial deposits.

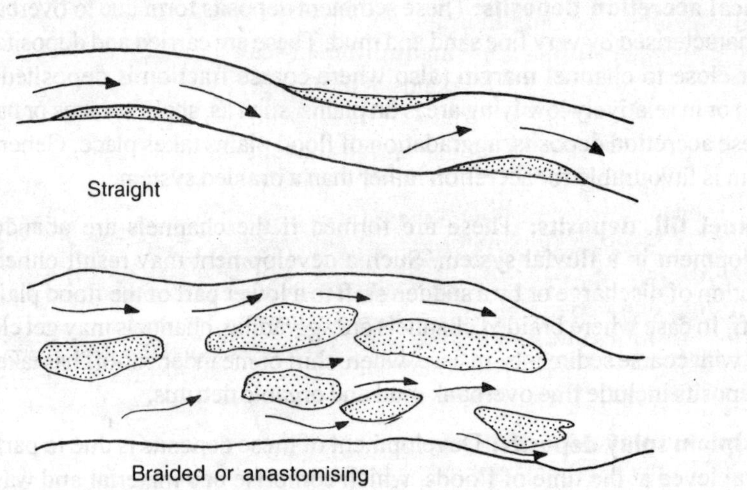

Fig. 8.5. Types of channel patterns.

Straight or meandering channels differ from each other, mainly in the context of the degree of sinuosity. Similarly, anastomising channels differ from braided in that the channel positions are relatively stable. Braided channels are characterised by higher slopes than meandering channels. Their formation is generally in response to high regional slopes, variable discharge and predominance of easily eroded, coarse bank deposits. In contrast, anastomising channels have low gradient. On the banks are cohesive sediments, formed out of the suspended load. Growth of vegetation further increases stability of the channels and bars.

Meandering Channel Model: Instability of flow in the channel seems to be responsible to give rise to meandering. A strong, spiral secondary flow, which results due to action of centrifugal forces, is responsible to give rise to characteristic features in the channel. The moving fluid particles are deflected towards the outer bank of a meander. This deflecting force is higher near the surface, where the velocity is high, than near the bed, where the velocity is low. Consequently, therefore, the secondary flow is spiral, with a reversal in its direction. Such a change takes place with every reversal in curvature of the meandering channel. Maximum velocity is not at the centre but at the outer bank of the channel. With the result, there is asymmetrical distribution of shear stress on the bottom and banks of a curved, meandering channel. Therefore, on the outer bank there is maximum erosion of the bend. Due to erosion, deepening of the channel and undermining of the outer bank take place. Sediments derived by erosion move downstream, across the channel by secondary flows near the bed or are diffused by turbulence as suspended load. These sediments, on deposition, give rise to point bar, in the inner part of the channel. The point bar deposits may or may not get exposed, which is related to the water level in the channel. At the higher water level, they are covered and with the lowering of water level, they get exposed. With lateral shifting of the channel, point bars accrete across the flow direction, giving rise to cross-bedded troughs.

Migration of point bars generally produces a characteristic fining upward sequence of four main zones which have been identified as:

(*i*) poorly bedded gravels, gradually grading into
(*ii*) trough cross—bedded coarse sands,
(*iii*) horizontally bedded medium sands and
(*iv*) ripple cross—laminated fine to medium sands

This sequence represents a simple meandering model.

Flow in channels is responsible for producing regular lateral and vertical changes in sediment characterstics and bed forms, while slow migration of the channel accordingly builds corresponding vertical sequences of sediment texture and structure. Similar fining upward sequence can also be produced by migration of meandering tidal channels.

Braided Channel Model: As mentioned earlier, braided channels are rather unstable and their movement or shifting often takes place. It has been observed that there are continuous changes and variations in the bed forms and sediment characteristics. Hence, construction of a comparative model is possible. There are two main types of braid bars, as; longitudinal braid bars and transverse braid bars. Longitudinal braid bars commonly develop in gravels and aggrade downstream. The initially deposited coarse sediments, which stream was unable to transport temporarily, result into longitudinal braid bars.

Gravelly or proximal braided systems are predominated by longitudinal bars and crudely developed horizontal stratification; whereas, in sandy or distal braided systems, transverse and diagonal bars develop. Dunes, ripple marks and sand waves are common features of the major channels. As the channels migrate, growth of bars takes place, which is characterised by alternating coarse and trough cross-bedded channel deposits, indicating cyclicity. These, in turn, are overlain by finer bar deposits, exhibiting planar cross-bedding, plane lamination and ripple cross-lamination. However, these cycles are not that regular, compared to fining upward sequences, encountered in meandering channel system.

3. Sandy Deserts

Deserts are the areas where evaporation far exceeds precipitation. The desert regions experience extremes of climate. There is very scanty and erratic rainfall. Arid and semi-arid climatic regions occupy nearly one third of the present land surface.

Three main types of environments within the desert have been identified, which are:

(*i*) alluvial fans and ephemeral streams;

(*ii*) inland sabkhas or playas and coastal sabkhas, and
(*iii*) sandy deserts, also termed as 'sand seas' or ergs.

Modern deserts seem to consist predominantly of eroding mountains (about 40%) and stony areas (about 10-20%), where erosion is more than the accumulation of sediments.

Studies indicate that ancient alluvial fan sediments are more in abundance than the ancient sand dunes, which is contrary to their occurrence in the present day environments. This suggests that the recent climatic history, especially that associated with the Pleistocene Glaciation and of later period, has favoured the formation of sandy deserts.

The largest desert of the world is Sahara, occupying an area of about 7 million km^2. It is composed of number of individual 'sand seas' or 'ergs'. Such individual bodies at times occupy an area as large as 5,00,000 km^2. Deposition of sandy deserts is generally favoured in structural and physiographic basins; having a long history of sedimentation. Most of the ergs appear to have deposited in topographic depressions, while the predominant wind direction seems to be crossing these sandy dunes.

Dune patterns in the Sahara desert are very complex, which is predominated by constant trade winds. Even large accumulations of sand dunes modify the wind pattern locally by topographic deflection and also due to extreme heating of sand surface that produces convective air circulation.

The eastern limit of the vast arid region of the Sahara desert is marked by the Thar desert. The eastern limit of the Thar terminates with the Aravalli hill ranges in India and its western limit merges with the fertile plains of the Indus river in Pakistan. Much of its area lies in Rajasthan State. Its total area is about 3.2 lakh km^2. It is dominantly sandy desert and is, therefore, termed as the Great Indian Sand Desert. Fluvial and aeolian processes operate in the desert. The fluvial action is dominant in the east; whereas, the aeolian action predominates in the western part of the desert. Sand dunes, sandy hummocks, sand streaks and sand sheets are the major features, which are produced due to aeolian processes. These bed forms cover more than 70% of the desert area. Isolated barchan with an average height of 2m to 8m. occur in the western part.

The Thar desert has been studied in detail from various aspects, like; climate, soil, geology, geomorphology, etc. These studies have been published in the Special Publication of Geological Society of India, Banglore (Singhvi and Kar (ed.),1992).

4. Lakes

Lakes are the standing water bodies, which are more commonly filled with fresh water. However, highly saline character of many lakes has also been reported. Lakes have been grouped and classified on the basis of various aspects; such as; its shape, length, breadth, depth, etc. Shape of a lake may keep on changing with time. A thick sequence of lake deposits may result under favourable conditions of sedimentation tectonics. A wide variety of geological and physiographic characteristics have been noted in a number of lakes. Largest lake on the earth's surface is the Caspian sea that covers an area of about 436,400 km^2. Another large sized lake is the Great Lakes that encompasses an area of about 245,240 km^2. There are various geological processes by which lakes originate. Also the physical, chemical and biological processes operating in them are much diverse. These facts have put constraints on generalised classification of lakes.

According to Reineck and Singh (1986), Reeves (1968) has distinguished following types of lake basins, as;

Structural basin, mass-movement basin, volcanic basin, meteoric basin, solution basin, glacial basin, fluvial basin, wind basin, organic basin, aeolian basin and shore line basin.

A major parameter controlling the characteristics of these lake deposits is climate. It controls the amount of precipitation and evaporation and therefore, nature of weathering and nature of soil in the source area. These aspects, in turn, affect the types of deposits in lakes. Climate is also responsible in seasonal fluctuations in lake levels. Rate of incoming clastic sediments and biota is also a function of climate.

Another scheme and the detailed classification of lakes has been suggested by Hutchinson (1957); as: (*i*) tectonic, (*ii*) volcanic, (*iii*) mass wasting (landslide), (*iv*) glacial, (*v*) chemical (solution action), (*vi*) fluvial, (*vii*) aeolian, (*viii*) shoreline, (*ix*) organic, (*x*) meteoric and (*xi*) anthropogenic (reservoirs, abandoned mine pits, etc.)

Whereas, Hakanson and Jansson (1983) have classified lakes on the basis of trophic status and water temperature (thermal). They have sub-classified trophic level lakes into: (*i*) oligotrophic, (*ii*) eutrophic and (*iii*) dystrophic.

Oligotrophic lakes have low primary productivity, high secchi-disc transparency, low algal volume, comparatively low concentration of nutrients and fish fauna. Eutrophic lakes have high productivity, low transparency and excess concentration of nutrients. Dystrophic lakes have higher amount of organic input, supplied from allochthonous source and low productivity.

On the basis of thermal characters, lakes are classified, as: (*i*) amictic, (*ii*) cold monomictic, (*iii*) dimictic, (*iv*) warm monomictic, (*v*) polymictic and (*vi*) oligomictic.

In amictic lakes, water is always frozen, while in cold monomictic lakes, temperature of water remains below +4°C and in such lakes, there is only one period of circulation of water in summer. Dimictic lakes are those in which water circulates twice a year. It also exhibits thermal stratification. In warm monomictic lakes, water circulates once a year, temperature of which never drops below +4°C. These lakes are characteristic of zones and mountainous subtropical areas; whereas in polymictic lakes, there is frequent circulation of water. Oligomictic lakes, however, are characterised by either irregular or no circulation and temperature in such lakes remains always above +4°C. These lakes are typical of tropical regions.

Selley (1976) has recognised four major categories of lakes. It is the aridity of climate and the supply of terrigenous sediments that decipher sedimentary facies of lakes. Accordingly, four categories have been identified as: (*i*) glacial lakes (*ii*) terrigenous permanent lakes, (*iii*) autochthonous lakes and (*iv*) ephemeral lakes.

Glacial lakes are those which are directly fed by glacial meltwaters. Terrigenous permanent lakes have low aridity and high inputs of terrigenous sediments. Their origin is related to glacial activity, but sediment input, however, is not controlled by glaciers. Lakes are characterised by laminated sediments, deposition of which is due to density currents. Autochthonous permanent lakes also have low aridity but low input of terrigenous sediments. The central part of these lakes may be rich in algally precipitated carbonates, copropelites or sapropelites. Ephimeral lakes are found in regions of arid climates. Common types are intermontane playa lakes, with mostly terrigenous muds and inland sabkhas (saline playas) and soda lakes with chemical precipitates.

Playas are generally flat, topographic basins of interior drainage, generally sarrounded by mountains and alluvial fans and are found in arid to semiarid climatic zones. Playa is flooded in full or in part from time to time to give rise to playa lakes. Though most playas have internal surface drainage, all of them do not necessarily accumulate evaporites, as the groundwater can escape from the basin, carrying dissolved salts; whereas, many playas do accumulate evaporites like that of Na, K, Ca and Mg carbonates, sulphates, nitrates, etc.

These schemes of classification clearly indicate that a very important parameter that strongly controls the lake deposits is climate. It controls the amount of precipitation and evaporation, nature of weathering and nature of soil in the catchment area and its vegetation. Considering these aspects, Reineck and Singh (1986) have proposed three major groups of lake deposits; as; (*i*) Classtic lake deposits, (*ii*) Chemical lake deposits and (*iii*) biogenic kae deposits. However, transition from one type of lake to another is possible.

(*i*) **Clastic lake deposits:** These are found in regions from arid climate to cold climatic zones near glaciers. Similar deposits are also encountered in sabkha, glacial lakes, ox-bow lakes and lagoons. According to Twenhofel (1932), an idealised distribution of sediments in lakes is characterised by another belt of beach pebbles , followed by a zone of sand, an inner zone of sandy, marly mud and mud in the central part. Such a zoning corresponds to the distribution of hydraulic energy, viz., the breaker zone, followed by a zone above wave base and a zone below wave-base. However, variations in distribution of sediments may result, which is due to prevailing wind direction, steep slope of the coast, nature and source of sediments. No vertical distribution of sediments can be constructed with the help of lateral distribution of sediments. This is because a lake does not grow from the margin.

(*ii*) **Chemical Lake Deposits:** These are very important in lake deposits and sometimes yield workable deposits of various salts. The formation of inorganic Ca-Mg carbonates and their diagenesis have also been reported from lake environment. In some lakes, enrichment of iron and manganese takes place, but with little economic significance. These deposits help for the discussion of chemical balances in lakes and related metal transfer from evenly stratified lake sediments and supernatant bottom lake waters. The chemical composition of lake water is controlled mainly by the dissolved and suspended material brought into the lake by a river and/or a brine. A number of salts are precipitated in a lake, as a result of chemical and biological processes. The evaporite deposits that form in a lacustrine environment have been reported and listed by Reineck and Singh (1986), as:

Carbonate Group: Limestone, aragonite, dolomite, magnesite, Mg-calcite, natron, trona.

Sulphates: Anhydrite, gypsum, glauberite, epsomite, thenardite, polyhalite.

Chlorides: Halite, sylvite.

Borates: Borax, searlesite, colemanite.

Nitrates: Soda nitre and nitre.

(*iii*) **Biogenic Lake Deposits:** Considering marine environment, very few species inhabit lake environment. There are a number of groups or species which never lived in lake environment. Generally, macrobenthonic fauna of lakes live in water depth less than 10 m, and a few macro and micro-organisms live at greater depths in lakes and only a few species of bacteria live abundantly at greater water depths. The sediments of the central basins are mostly free of any bioturbation structures.

The plant life is commonly abundant in shallow parts of the lake bottom and on the water surface. As the lake is getting filled, piants become important in the last stage that give rise to different kinds of plant deposits. Several plants living in lakes produce carbonates. During their decay, they produce lake marl. The bicarbonate content of lake water in glacial lakes is derived from weathering and leaching of glacial till, during the post-glacial period, Lamellibranchs and gastropods also contribute to the carbonate content of lake sediments. In present days, carbonate-rich lake environments result due to low precipitation-evaporation ratio. These carbonates are generally aragonite, high Mg-calcite, depending upon the progress of carbonate precipitation, governed by Mg/Ca ratio of lake water. Lastly, dolomite may be formed in a fresh water lake environment.

In eutrophic lakes, characterised by high production of nutrients and low oxygen content, decay of plants leads to humus-gel formation that ultimately converts into humus coal, e.g. cannel coal. In deeper parts of some lakes, black sapropel sediments are deposited, which are very rich in organic matter. These sediments may develop into oil shale. Shallower is the lake, broader is the plant inhabitated zone near the shore and death and burial of plant population produces deposits like peat. Clastic lake deposits are also strongly associated with lake marl and sequence of sedimentation may terminate with peat or coal deposits.

5. Coasts (non-deltaic)

The Pleistocene glaciation has brought out variation in sea level, resulting into transgression - regression cycles on all the coastlines of the world. On the basis of morphologic features and the way the coastlines have evolved, Fairbridge and Finkl (1988) have classified the world's coastlines into three types; as; emerging, submerging and quasi-stable. Each has its own characteristic assemblage of geomorphic features. Similarly, on the basis of tidal range, coastlines may be classified into three types, as; macrotidal (>4 m), mesotidal (2-4 m) and microtidal (<2 m). Each type is characterised by distinctive depositional forms and facies. The major difference, however, lies between macrotidal and other coastlines. Meso and microtidal coastlines generally are characterised by barrier islands and lagoons, while in macrotidal coastlines, these features are generally absent. In macrotidal coastlines, waves play a significant role in carrying out coastal erosion and longshore transport of sand and gravel.

Geological setting, climate and changes in sea level are the major and important controls in deciphering development of facies along coasts. A typical macrotidal coast is characterised by funnel-shaped estuaries or bays, which contain large intertidal and subtidal sand ridges. These may range even upto a kilometer wide and may run for several kilometers. Such sand ridges are elongate and roughly parallel to the tidal currents and are separated by major tidal channels.

The features that distinguish macrotidal deposits from those of meso- or microtidal deposits have been summarised by Blatt, *et al.* (1980), as:

(*i*) total absence of the types of deposits, associated with coastal barriers;

(*ii*) large thickness of cross-bedded sands that result due to migration of megaripples and megarippled sand waves on subtidal to intertidal sand ridges. Palaeocurrents might be locally unimodal with cross-bedding interrupted by reactivation surfaces; or bimodal (herringbone type). However, generally it should be bimodal with the mode directing onshore.

(*iii*) scarcity of low-angled lamination formed by wave swash, except for minor occurrence in the outer zone sands;

(*iv*) fining upward sequence formed due to both progradation (ie. regression) and transgression in the macrotidal environment, in contrast to coarsening upward sequence, typical of prograding barriers.

In meso- and microtidal coasts, having low relief, development of barrier islands and shallow lagoons behind the islands takes place. An ideal barrier island model contributes the barrier island itself, with its shoreface, foreshore, ie. beach; backshore and dune environment. Lagoons behind the barriers are connected to the open sea by tidal inlets, which may migrate downdrift due to erosion by tidal currents. Deposition of sand on spits results due to wave-induced longshore currents. Occasionally, major storm may break barriers and rarely, a new inlet may also be formed. However, generally a broad shallow channel is cut through the dunes, barrier is eroded, resulting in the transport of sediments shorewards into the lagoon, as a washover fan. Lagoon also gets filled due to sediments brought by deltas from inland, alongwith the deposition of mud, brought by tidal currents from sea on the intertidal mud flats and supratidal salt marshes (Fig. 8.6).

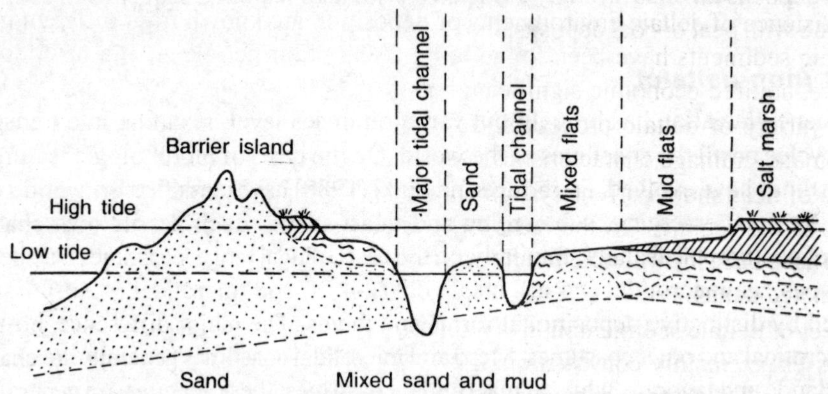

Fig. 8.6. Coastal environments.

Development of these coastal environments is generally controlled by the tidal range, rate of sediment supply, intensity of wave action and rate of sea-level changes.

Coastal dunes are formed due to blown off sand from adjoining beaches. Generally, these consist of fine grained sands than the beaches. Coarser fraction from the beaches is rather absent. Most of the dunes are stabilised due to growth of vegetation, and trapping of sand produces low-angle cross-bedding (10°-15°) alongwith high angle cross-bedding (25°-35°), formed on the slip face of migrating dunes. Orientation of low angle cross-beds is found to be in any direction. In fact, it is polymodal; while high angle cross-bedding in most of the coastal dunes is observed to be bimodal and the bisecting direction is in the direction of the prominent sand-transporting wind. Tidal inlets may be relatively stable or may migrate rapidly. Filling of old inlets and opening of new ones also take place.

Transgression and Regression: It has been noted that majority of barriers have been built up during the rapid post-glacial rise in sea level, i.e. about 17,000 to 5000 years B.P., and later followed by a slow rise in sea level or stable sea level. Sea level rise has played a significant role in origin and modification of recent barriers. However, such barriers are not necessarily the result of sea level rise. The main factor that is involved in transgression-regression in such cases appear to be the balance between transport and supply of sediments.

Considering other factors being constant, a general response to sea level rise is coastal erosion from the foreshore and upper shoreface and corresponding deposition of the eroded material in the lower shoreface and offshore regions. It is so observed even in case of slight rise in sea level.

Another possible response to rapid rise in sea level is drowning of coastal barriers, resulting at least in partial preservation of barrier sands in the form of relict, linear sand accumulation on the continental shelf. Such 'relict' sands on the continental shelf are found to be reworked slowly by waves or currents. However, for rapidly rising sea-level, drowning of sand barriers or even shoreface retreat may result. Onshore-offshore movement of sand takes place mainly alongshore. Formation of sandy or muddy wastes mainly depend on relative magnitudes of coastal wave energy and rate of sediment supply.

6. Delta

Deltas are the complex depositional systems and are the major sites of accumulation of terrigenous sediments. Existence of deltaic environments of deposition are known from different geologic periods. Deltaic sediments have been found to be the hosts of petroleum and natural gas and have, therefore, aquired economic significance.

Due to a variety of deltaic processes, different delta types result. These different types depend on climate, drainage characteristics, discharge of water, sediment yield, tidal and wave energy, nature of near shore currents, continental shelf-slope on which the delta is being built and tectonics of the regime. On the basis of these different controlling factors, three major types of deltas have been identified, as: (*i*) fluvial dominated deltas, (*ii*) wave-dominated deltas and (*iii*) tide dominated deltas.

Major aspect of deltaic sedimentation is interplay of constructive and destructive processes. Fluvial processes are mainly constructional, in which delta builds out into standing water by extension of major distributory channels. In such a process, offshore deposition of mud from the mouth of the major distributory channel takes place, preceded by sand deposition at the mouth of major distributary channel. The delta framework so established, gradually grows by the processes of crevasse and splay formation and by redistribution of sediments by waves and tides. In fluvial dominated delta, constructional phases give rise to most of the features of an active lobe, which then are termed as high-constructive deltas. Shifting of water and sediment discharge to a new distributory channel as a constructive phase may rework the earlier, abundaned channel. In case of high-destructional deltas, sediments supplied through major channel are constantly redistributed by marine processes. In such cases, therefore, constructional and destructional phases are not that apparent.

Major processes that operate in a deltaic environment have been discussed in the following paragraphs:

In river-dominated deltas, river waters leave the distributory mouths and enter into still waters of depositional basins. It is speedy, moving water that enters into the basin waters. When river water enters the sea, it has a lower density than sea water, thereby allowing suspensoids to remain in river water. Due to the density difference, therefore, river water spreads like a plane jet above sea water. Mixing of water is rather slight and jet spreads out slowly. Suspended sediments, therefore, are carried out to a certain distance from the mouth, and tend to get deposited in two areas, viz., in subaqueous levees and distributory mouth bars (Fig. 8.7), where speed is gradually reduced due to mixing. Store currents, not related to river processes, carry finer sediments farther and farther to be deposited on a wider area to give rise to prodelta muds.

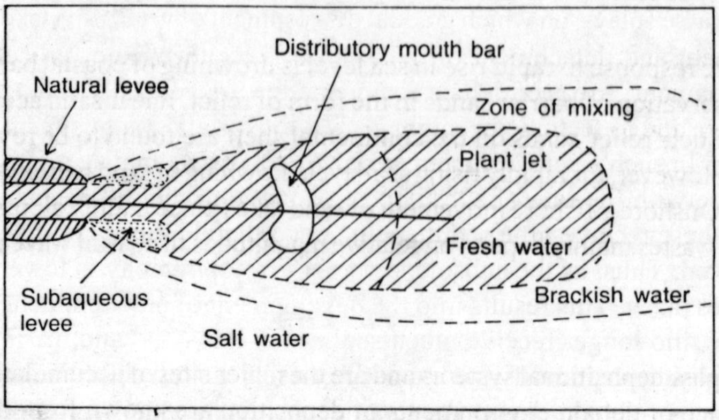

Fig. 8.7. Distributory mouth bar of a delta.

As river water enters sea water relatively with some higher speed, it tends to give rise to reverse flow of sea-water, or salt water, in the form of "wedge" that intrudes the river channel and travels upstream, in some cases, even for a considerable distance. Sediments get mixed up at the interface of salt water, giving rise to a slurry-like mixture.

Salt water intrusion is associated with buoyant force, and is responsible for the formation of a plane jet. The intrusion may get reduced, if the main distributory fluvial channel is discharging into shallow rather than deep water. Due to shallow nature, therefore, there is a sharp reduction in jet flow, which then gives rise to a large, shallow, middle ground bar.

In many deltas, processes at the river mouth get strongly modified due to waves and tides. If the tides are large, velocities of the fluvial-distributory channels are either checked or even reversed, producing strong seaward velocities of ebb tides. Such a process widens the river mouth. If the tidal range is more than 4 m., funnel shaped estuary is formed and sand accumulation results in the form of elongate tidal ridges. The flood tidal waves that travel upstream become asymmetrical, produce short duration currents, distributary sand into tributary tidal channels and on to tidal flats. Accumulation of sand at the distributary mouth is prohibited by strong wave action. However, when large amount of sand is supplied through distributary channels during floods, sand gets redistributed by longshore currents to give rise to beach ridges.

Formation of crevasse splay deposits in deltaic environment is another major process. The low gradient is typical of many deltaic plains; and which are prone to overbank flooding during periods of high discharge. At these times, some of the coarser suspended sediments are deposited along the distributary banks to form levees, which in turn, restrict the river flow to the channels and reduce the frequency of flooding of interdistributary areas. However, at the time of high flooding, levees may get locally washed away, forming shallow channels (i.e. crevasse), and giving way to substantial amount of flow. Such crevasse may be closed by subsequent sand deposition to give rise again to levees. A few may develop in distributary channels, branching off from the main channel.

In river-dominated deltas, crevassing plays a significant role in forming subaerial deltaic plain. Initially, as the distributaries develop, primary delta framework gets constructed. Areas in between the distributaries are occupied by shallow bays or freshwater lakes. Later, these are filled by sediments, as crevasse splays develop. These crevasse splays are initially small lobate deltas, which are formed within the larger framework of delta. Overbank flooding fills the interdistributary areas and crevasse splays, on which gradual development of vegetation takes place.

Development and detachment of delta lobes is another major process that is operative in deltaic environment. Areas of deltaic sedimentation are generally characterised by subsidence, which is either due to tectonic or might be due to isostatic readjustment in response to deltaic sedimentation.In such an environment, sediment lobes accumulate around active distributaries, that keep pace with the rate of subsidence. Along with the distributaries, adjacent areas also subside, but these receive little sedimentation and therefore possibility of transgression by sea water is created. Thus, as the subsidence proceeds, topography is lowered and shift in fluvial channels takes place. This results into the development of new sediment lobes. The earlier, old sediment lobes no longer receive much sediments. Subaerial and marine erosional processes, along with subsidence, modify the exposed parts of the old delta.

Deltas which are not strongly river-dominated, displacement of sediment lobes is not much important. In some cases, position of the main distributaries are relatively much stable; whereas, in some other deltas, sediments delivered through the main distributary are spread over the entire delta front by wave and tidal action. In such cases, displacement of delta may not take place or it may occur less frequently.

Displacement of delta, associated with subsidence may give rise to alternating marine and non-marine sequences, with much of the characteristics, similar to clastic cyclothems of coal measures. Formation of such sequences can be explained more plausibly by steady subsidence, associated with displacement of delta. Presence of marine sequences suggests transgression mainly due to eustatic sea-level rise.

7. Terrigenous Shelf Environment

It has been observed that sedimentation on present day continental shelves is mainly affected by post Pleistocene sea-level rise. The sediment and other lithological characteristics lead to suggest that majority of 'sand' on modern shelves are relict sands; deposited when sea-level was low. However, Swift, *et.al.*(1971) are of the opinion that eventhough majority of sediments have been deposited on continental shelves at the time of low sea-level stands, these are now being reworked by the present,current processes, which are therefore, modifying shelf morphology and characteristics of older, relict sediments. These sediments have, therefore, been termed as palimpsest rather than relict. To determine the nature of currently active processes on continental shelves is rather a difficult task. It is because the continental shelves were and have been subjected to sea-level changes during geologic past. Even the extent to which the present sea-level is to remain stable to affect sedimentation on continental shelves cannot be accurately estimated. From the nature of processes, operating on modern shelves, two major types can be identified, as (i) tide dominated, and (ii) dominated by waves and marine currents.

Generally, continental shelves are subjected to the action of waves and tides. Waves produced by storms carry sediments to considerable depth. But if the tides are large, tidal currents are responsible for sediment movement. Bed forms, produced due to tidal currents, get modified later by wave action, particularly in shallow water.

Belderson and Stride (1966) (in Blatt, *et.al.*, 1978) recognised five different facies of terrigenous shelf environment as

1. gravel lag—areas of erosion;

2. sand ribbons, which are few cm. thick, about a meter wide and elongate and parallel to tidal currents.

These two facies are formed in an area which is swept free of sand or bypassed by sand.

3. sand waves, ranging from 1 to 10 meter in height with wavelengths of 200 to 500 meter. Crests orient normal to tidal currents and vary in form from symmetrical to strongly asymmetrical; smaller megaripples are generally superimposed on sand waves. This facies develops in the main zone of accumulation;
4. rippled fine sand and muddy sand;
5. sand patches, with intervening shell gravels.

The last two facies develop generally in such areas, where sand supply is less and currents are weaker.

Topographic depressions, areas of low tidal currents and those which are closer to major source of mud are the sites of mud accumulation.

Some modern continental shelves are characterised by waves and currents that are generated by wind. Shelves are marked by broad valleys, which are produced due to retreat of estuarine mouth. These valleys may not coincide with burried channels of rivers, that flowed at the time, when sea level was lower. Littoral drift of sediments develops large, low ridges, and when superimposed on large scale topography, a system of ridges is produced. These ridges are tens of kilometers long, about 2 to 5 km. apart and are maximum upto 5 meter in height. Some ridges are produced mainly at the foot of modern shoresurface; which may get detached by subsequent transgression. These ridges also get continuously modified due to the development of closely spaced secondary ridges and troughs. Sand waves may get superimposed on ridges and finer sand accumulates in troughs. Oscillating and unidirectional currents produced by storms are responsible for sediment movement. Sediment supply is from coastal and shelf bottom erosion, that produces transport of sediments and sediment fining towards the shelf edge.

Development of three main facies on continental shelf generally takes place, which are mainly (i) sand, (ii) mixed sand-mud, and (iii) mud.

8. Carbonate Shelf

Carbonate sedimentation in shallow-water environment takes place on platforms of two types. In first case, the platform is attached to land masses, while the second type is isolated shallow water bank that rise from oceanic depths, or the coral atolls. Examples of the first type is the Persian Gulf, whereas that of the second one is Bahama Platform, both of which have been extensively studied and cited in detail in the literature.

Pre-requisite for the formation of carbonate platform is extensive growth of carbonate producing organisms. For the same, shallow-water environment is required, wherein light-requiring organisms can thrive. In such an environment, little influx of terrigenous sediments dilute carbonate contribution to give rise to impure carbonate sediments. Introduction of terrigenous silicate sediments in carbonate sediments produces muddy water, affecting the depth of penetration of light, thereby creating inhospitable environment for the survival of organic life. Addition and mixing of silicate terrigenous sediments hamper the production and deposition of carbonate sediments. However, case of isolated platform that rises from oceanic depth, elimination of terrigenous clastic sediments easily takes place. Detrital terrigenous material may enter oceanic depths, and its transport upon the platform is rather difficult.

Once the formation of carbonate platform has begun, thick accumulation of carbonate sediments is possible, if the shallow water environment is maintained by gradual subsidence. Due to subsidence, slightly deeper water becomes still deeper and new carbonate sediments deposit on previously deposited carbonate sediments in shallow water environment. With the result, platform having steep sides and abrupt margins is formed. Changes in sea-level further bring out variations in carbonate sedimentation.

B. CONTINENTAL MARGIN

9. Continental Slope, Rises and Basin Plain

Boundry between continental shelf and continental slope is defined by distinct break in slope. This break generally takes place at a distance of about 75 km. from the coastline and at a depth of about 130 meters, commonly with a range of 20 to 200 meter or more. Continental slope is relatively narrow, having a width of about 20 to 100 km. and extends upto depths of 1500 to 3500 meter. Its slope varies between 1° to more than 20°, averaging to about 4°. Base of continental slope is marked by a broad region, having somewhat lower slope, termed as continental rise. It has a width upto 600 km., with a gradient generally less than half a degree. Continental rise, in turn, gives way to abyssal plains or basin floor with very low relief (Fig. 8.2).

Width of the continental slope off Kutch, NW of India is minimum, and is about 19 km., slope edge being at a depth of about 1450 m. At the base of the slope, several depressions and elevations have been noted. General relief is of the order of 100 m., occasionally rising to 300 m. There is a marked increase in the width of the slope in the southern direction. It is about 39 km. off Dwarka, about 50 km. off Porbandar, increasing to about 84 km. off Bombay. Regional gradient varies between 1 : 9 to about 1 : 25-1 : 35.

The continental shelf and slope together constitute what is termed as 'continental terrace', (Fig. 8.2). These are more commonly the sites of deposition rather than erosion. Broad continental shelves, slopes and rises are generally the characteristics of most stable, i.e. intraplate or trailing continental margin, which is formed due to extending wedge of terrigenous and carbonate sediments, in response to gradual sinking of continental margin.

Progradation of continental shelf margin is mainly due to the supply of terrigenous and biochemical sediments. Continental slopes, in turn, are built by the sediments that are worked over from the continental shelves.

Gradient of the continental slopes averages 4°. If it exceeds to 10° or more, it is generally due to faulting or by structural disturbances and also due to build-up of coral reefs. Generally, low gradient of the slope is maintained either due to slumping or development and formation of syndepositional growth faults.

Another prominent feature of the continental slope is presence of submarine canyons. These are generally of great length and depth. Starting from the base of the continental slope, canyons extend as a submarine fan for more than 300 km. to a maximum depth of about 4500 m. Canyons are either related to rivers or head into the near shore zone.

Submarine canyons are generally characterised by the presence of submarine fans, below which are the sites of sediment accumulation, like alluvial fans. The Bengal fan is the world's largest fan, which is about 3000 km. across, even crossing the equator, in the Bay of Bengal and is fed by sediments supplied through Ganges-Brahmputra delta, to a canyon that cuts across the continental shelf edge.

Deep sea basins considerably vary in size. Continental margins are marked by small basins, which are broken by faulting, or by the presence of small ocean basins, whereas, large abyssal plains are several hundreds of kilometers across and are typical of large ocean basins. Abyssal plains which receive terrigenous sediments mainly due to turbidity currents, and are characterised by very low relief, with a gradient generally less than 0.001.

10. Subducting Plate Margins

The environments described in the previous paragraphs are from the region of tectonic stability. The same environments, i.e. continental shelves, slopes and basins are also encountered in tectonically active areas; such as, the areas of convergent plate boundaries. These environments have been identified as oceanic trenches, slope basins, fore-arc basins and marginal basins (Fig. 8.1).

Subducting plate margins are basically of two types, *i.* convergence of two oceanic plates, which is characterised by island-arc-trench margin, and *ii.* convergence between continental and oceanic plates, characterised by mountain arc-trench systems. In both the cases, there is a wide gap between the trench and the island arc or mountain arc, which is termed as the arc-trench gap. Associated environments with subducting plate margins are shown in Fig. 8.1.

C. OCEANIC FACIES

11. Oceanic Trenches and Slope Basins

Oceanic trenches are tectonic trenches and are the deepest in oceanic environment. These are generally filled with sediments, which are either terrigenous or volcanogenic. Majority of the trenches in the Pacific ocean have nearly 400 m. thick sediments, which are pelagic sediments overlying the oceanic crust, followed by hemipelagic sediments, represented by mud and some amounts of turbidites. Arc side of the trench is steeper than the ocean side. Therefore, slumping, sliding of the arc side of the trench is a common observation. Even canyons and submarine fans extend into the trench, giving rise to thick sequences of sediments that overlie abyssal pelagic sediments.

It is also a possibility that sediments derived due to erosion of islands or mountain arcs get trapped either in 'fore-arc' basins or in basins formed on the arc side slope of the trench. It is therefore, possible that pelagic sediments may subduct with the volcanics, while hemipelagic mud and turbidites deposited in a trench are thrusted up during subduction, forming the arc side of the trench. Thrusting up of the sediments temporarily give rise to small basins, which are short-lived but even during the course, they entrap the sediments.

Fore-arc Basins: Basins that are formed in the arc-trench gap are underlain either by oceanic crust or by thin continental crust. These receive sediments due to erosion of the adjacent islands or mountain-arcs, giving rise to thick sequences. These sediments have a fair chance of preservation rather than those deposited in trenches; possibly because of relatively more stability than trenches.

All the fore-arc basins are not necessarily steepThese basins are generally underlain by thin continental crust and receive sediments, which are either fluvial or deltaic or of shallow marine facies. Due to tectonic instability, however, there is a possibility of rapid changes in facies associations.

Marginal Basins: The sea between a continent and an island-arc gives rise to marginal basin. It is a structural basin, underlain by oceanic crust. Some basins are characterised by active spreading, and presence of thin sediment cover; whereas, in inactive basins, continued sedimentation and subsidence give rise to thick sequences of sediments.

Most marginal basins are generally cut off on the arc side and therefore there is a paucity of terrigenous sediments of continental origin. Sediments in these basins are mainly pyroclastic and volcanic sandstones, source of which is the volcanic arc. Alongwith these sediments are also present windblown dust derived from the adjacent continents and also the biogenic sediments.

12. Oceanic Ridges and Basins

According to the concept of Plate Tectonics, it is believed that pelagic sediments are subducted alongwith the oceanic crust. The same concept also believes that some remnants of oceanic crust (i.e. ophiolites) are thrusted up, rather it is obducted on the ontinent.

Present distribution pattern of pelagic sediments suggests that these sediments accumulate on the ocean floor, as it moves away from the spreading centres, i.e. mid-oceanic ridges. As the subsidence of the oceanic crust takes place, it moves from above the carbonate compensation depth, where carbonate oozes accumulate, into the region below the carbonate compensation depth, where siliceous oozes accumulate. This results into vertical sequence of pelagic sediments from carbonate to non-carbonate, followed by hemipelagic or volcanogenic sediments. Such a sequence of sediments, however, gets disturbed either due to irregular topography of the ocean floor near the ocean ridges or by relatively strong ocean currents, present in deep ocean basins. These sediments are generally devoid of terrigenous clastic sediments. However, prograding clastic wedges from continental shelf and slope through submarine fans may enter into the pelagic and hemipelagic sediments at the ocean ridges and basins.

9

CHAPTER

Allochthonous Rocks: Sandstones

9.1 NOMENCLATURE AND CLASSIFICATION

'Sandstone' is an important class of sediments and they contribute to our understanding of geologic history more than any other sediments. Composition of sandstone is a clue to provenance. Their directional sedimentary structures yield an information about palaeocurrent, while their geometry alongwith internal structures throw light on environment of deposition.

Sands can be divided into three major groups, as; terrigenous, carbonate and pyroclastic. The terrigenous sediments are largely composed of fragments of clasts, which are derived from earlier pre-existing igneous, sedimentary or metamorphic rocks. The clastic grains are produced by mechanical and chemical weathering and are then transported, sorted, modified in size and shape, before finally reaching to a site of deposition.

Carbonate sands are mostly marine and are primarily skeletal grains, oolites or intraclasts (intraformational particles). These sands are produced within the basin of deposition and are not the fragments or clasts derived from pre-existing rocks.

Pyroclastic sands are produced by volcanic explosions. They may get transported through air or in water. The term volcaniclastic is also used to sands, which are rich in volcanic debris that are either truly pyroclastic or terrigenous derived from an older volcanic deposit or formation.

The terrigenous clastic sediments are a diverse group of rocks, that range from fine grained mudrocks through coarse grained sandstone to very coarse conglomerates and breccias. The detrital grains in terrigenous sediments may be rock fragments, but majority of the grains may be individual crystal or mineral, chiefly of quartz and feldspar, that are abraded to various degrees. A very fine break-down product of the source rocks, which is formed due to weathering process, consists mainly of clay minerals and are predominant in mudrocks. Such a fine product may also form the matrix of some conglomerates and sandstones.

Understanding of sandstone has been increased through studies of modern sands, by knowing their manner of formation and their accumulation. In the present day environment, most common sites of sand accumulation are linear, such as; rivers and beaches. However, majority of sands in the geological past occur as extensive, stratified deposits. Comparison between the essentially linear occurrences of present day sand deposition and extensive sheet like accumulation in the geological past lead to suggest that the latter happen to be the result of displacement of linear bodies of sand deposition or by lateral migration of streams or rivers or by transgression/regression of shoreline. However, some shelf areas where extensive sand deposition has taken place in present day seem to be exceptions.

Mineralogical and chemical composition of modern sands vary widely. However, the same depend more or less on size and lithological characters of the source area, compared to climate, environment and agent of deposition. The constituents of majority of sands are represented by quartz, felspar and rock fragments.

The most important and basic descriptive textural attribute of any sedimentary rock is its grain size. To classify the rock, however, the average grain size and distribution of grain sizes in a rock are given due consideration. For the purpose, several grain size scales have been proposed and grain diameter is considered as an adequate measure of size. The details of the grade scales have been presented in chapter IV. Various terms to indicate size are used, and are presented in Table 9.1.

TABLE 9.1: Grade scales, after Udden-Wentworth and phi scale for size analysis of sediments and sedimentary rocks; alongwith corresponding sieve numbers

mm	*phi*	*Class term*
4	−2.00	
2.83	−1.50	Granules
2.00	−1.00	
1.41	−1.50	Very coarse sand
1.00	0.00	
0.707	0.50	Coarse sand
0.50	1.00	
0.354	1.50	Medium sand
0.250	2.00	
0.177	2.50	Fine sand
0.125	3.00	
0.088	3.50	Very fine sand
0.063	4.00	
0.044	4.50	
0.031	5.00	
0.0156	6.00	Silt
0.0078	7.00	
0.0039	8.00	

In recent years, number of attempts have been made in suggesting various schemes of classification of sandstone. The first serious attempt in classification of sedimentary rocks was that of Grabau (1904) and later by Krynine (1948), Pettijohn (1948, 1950), Shrock (1946,1948), Klein (1963), Okada (1971) and Pettijohn, Potter and Siever (1972).

It can be seen that in most schemes of classification, sandstones are best described and classified on the basis of their texture and mineral composition. Of these, the classifications suggested on the basis of mineral composition prove to be most significant and meaningful; probably with few exceptions. In these systems of classification, definite distinction has been made between detrital components, constituting the framework of the sandstone and the cementing minerals as the bases for classification. It is seen that compositionally, sandstone has three principal constituents, as; quartz, feldspar and rock fragments. The latter constitute the sand sized particles of igneous or sedimentary or metamorphic origin. Other detrital components are generally rare. In exceptional cases, for example, glauconite can be a major constituent as in greensand or magnetite can be a major constituent as in black sand. The framework composition of sandstone, therefore, can be expressed in terms of the relative abundances of three major constituents, as quartz, feldspar and rock fragments. In different schemes of classification, therefore, proportion of each is shown graphically with the help of equilateral triangle. Major sandstone classes or families are therefore, defined on the basis of these three major constituent minerals. An equilateral triangle is further subdivided, as necessary, to define further sandstone classes or families.

These systems of classification take into account normal sands and also those having a well-defined framework. However, they enter into difficulty, where there is an appreciable amount of matrix, such as; wackes. Such a difficulty has been overcome by many, by suggesting two broader groups initially; as those with matrix and those without matrix and then to subdivide each group.

Such an approach in sandstone classification has been found to be workable and meaningful. However, in different schemes of sandstone classification, placing of partitioning boundaries and sandstone terminology have varied from one classification to another.

Various schemes of classification that are commonly used have been presented in Fig. 7.2.

In these schemes of classification, the mineral parameters form the basis of classification and prove to describe and categorise sandstone successfully. At the same, they also have the genetic significance.

These mineral parameters help to determine provenance factor and also maturity of sandstone. The ratio feldspar/rock fragment yields an information about provenance and distinguishes between deep-seated provenance and supracrustal provenance. Supracrustal rocks are either igneous, sedimentary or metamorphic and therefore should be fine-grained and yield sand-sized rock particles. The ratio of quartz/feldspar or quartz + chert/feldspar + rock fragment is an approximation of compositional maturity of sandstone. The ultimate end type is represented mainly by pure quartz. Another important factor to describe the characteristics of sandstone is textural maturity, which is described by sorting i.e. standard deviation and rounding of sediments. Fluidity factor is another parameter, which is rather difficult to determine and is expressed by the ratio, quartz + feldspar + rock fragment/matrix. Sediments with a high proportion of matrix are most likely the products of a quasi liquid or mass flow of mud-sand mixture. It is, therefore, considered as an index of fluidity. However, it is also likely that some matrix might be post-depositional, i.e. diagnetic. These are, therefore, the limitations in the interpretation of fluidity factor.

In the schemes of classification of sandstone, suggested by Pettijohn (Table 9.2), sandstones with 15% or more proportion of matrix are termed as wackes; and those with less are the ordinary orthosandstones. Thus, in the class of matrix free or matrix poor sandstones, three main types are defined:

1. sandstone in which quartz constitutes 95% or more of the framework grains is the quartz arenite;
2. sandstones which contain 25% or more of the feldspar and which exceeds rock fragments, are termed as arkoses; and
3. sandstones which are characterised by more than 25% of the rock fragments, are termed as lithic arenites or lithic sandstones.

These are further divided into subclasses, classes which are transitional between these important families.

Table 9.2: Scheme of classification of sandstone (after Pettijohn, 1957) *

<table>
<tr><th colspan="2">Cement or matrix</th><th colspan="2">Detrital Matrix Prominent (over 15%) to predominant Chemical Cement Absent</th><th colspan="4">Detrital Matrix absent or Scanty (under 15%) Voids Empty or Filled with Chemical Cement</th></tr>
<tr><td rowspan="5">Sand or Detrital Fraction</td><td rowspan="2">Feldspar Exceeds Rock Fragments</td><td rowspan="4">Graywackes</td><td rowspan="2">Feldspathic graywacke</td><td colspan="2">Arkosic sandstones</td><td rowspan="4">Orthoquartzites</td><td rowspan="2">Chert < 5%</td></tr>
<tr><td>Arkose</td><td>Subarkose or feldspathic sandstone</td></tr>
<tr><td rowspan="2">Rock Fragments Exceed Feldspar</td><td rowspan="2">Lithic graywacke</td><td colspan="2">Lithic Sandstones</td><td rowspan="2">Chert > 5%</td></tr>
<tr><td>Subgraywacke</td><td>Protoquartzites</td></tr>
<tr><td>Quartz Content</td><td colspan="2">Variable; generally < 75%</td><td>< 75%</td><td>> 75% < 95%</td><td colspan="2">> 95%</td></tr>
</table>

* Volcanic tuffs and intraformational calcarenites not included

As such, a scheme of classification is based on mineral composition. It is,therefore, independent of the depositional environment. Mineral composiion, however, is a reflection of the source rock characteristics, i.e. provenance and hence this classification, more closely reflects the source area rather than anything else. It is also indirectly related to tectonics.

With the advent of plate tectonics concept, different sites of depositional basins have been identified; and locations of such depositional sites have been changed through geologic time, as the plates move across the earth's surface. As the plates collide or diverge or slide past each other, different types of plate tectonics settings result, and which are accordingly associated with the mountain-building orogenic processes, metamorphism or magmatic activity. By studying the mineral composition of sandstone, therefore, attempts have been made to interpret types of source rocks and also the type of plate interaction that was responsible for the entire process. Such a relationship between plate movements and sandstone petrology has been first documented by Dickinson and Suczee (1979) and is presented in Fig. 9.1.

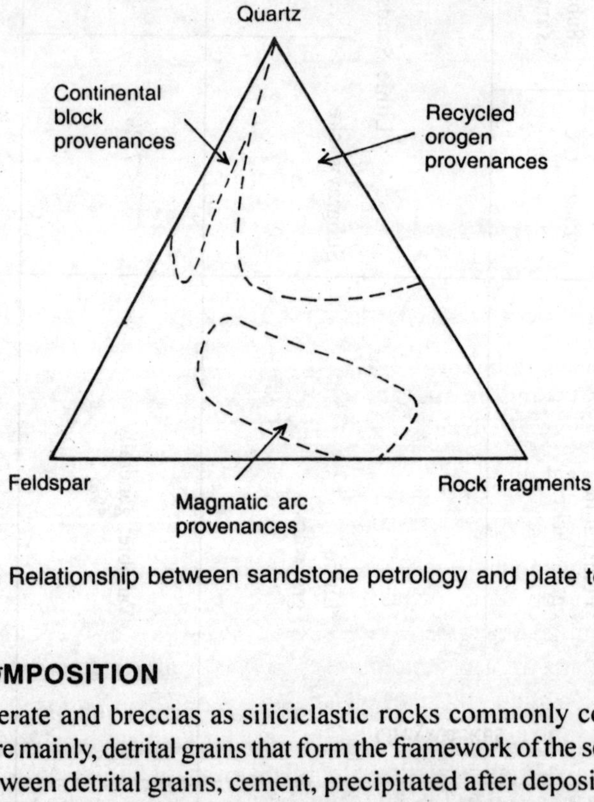

Fig. 9.1. Relationship between sandstone petrology and plate tectonics.

9.2 MINERAL COMPOSITION

Sandstone, conglomerate and breccias as siliciclastic rocks commonly consist of three major components, which are mainly, detrital grains that form the framework of the sediments, fine grained matrix that occur between detrital grains, cement, precipitated after deposition of the sediments and authigenic minerals formed during diagenesis. Such a categorization and assignment of minerals accordingly is necessary and meaningful mainly to interpret the history of sediments.

A mineral in its natural state and a rock fragment may occur in sediments, if the source rock had been subjected to incomplete weathering and transportation was too short. A number of detrital minerals, therefore, can occur in sediments. However though, in practice, only few species are encountered in sediments. Of these, quartz is the most dominant mineral species in sandstone. In some cases, it forms over 90% of the detrital fraction. Feldspar is another common mineral in sandstone, but it is present in very subordinate amount. Rock fragments are also present in some sandstones.

Quartz: It is the chief constituent mineral commonly occurring in 2/3rd of the proportions of sandstone. It is primarily a detrital mineral, while in some cases, it is also authigenic, i.e. occurring as overgrowth on detrital quartz grains (Photo 9.1). It is, therefore, an important cementing material in such sandstones.

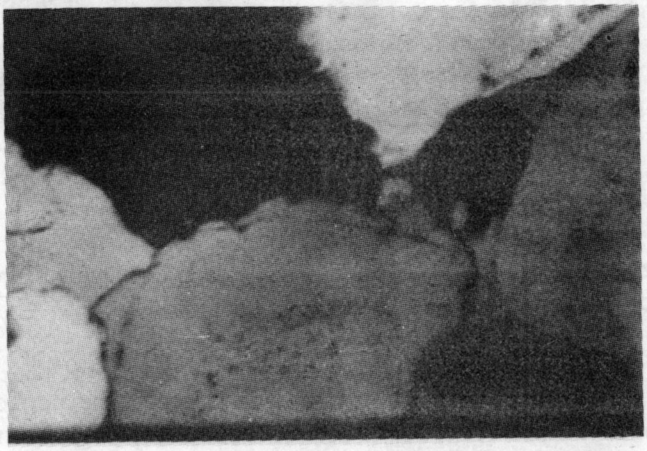

Photo 9.1. Overgrowths on detrital quartz grains, between crossed nicols, X55.

Detrital quartz usually has diameter less than 1.0 mm., more commonly less than 0.6 mm. Grains larger than 1.0 mm in size are either composite or polycrystalline. Nature and size of quartz grains are generally deciphered by the size of the quartz grains in the source rock. For the detrital quartz, the principal ultimate source rock is quartz bearing plutonic rock, namely granite.

The shapes of quartz grains also vary greatly. More commonly, they are subspherical, but show a slight elongation, rather in the direction of crystallographic axis. It is observed that quartz grains from igneous and metamorphic rocks and also those from granites tend to be elongated, preferably in the direction of c-crystallographic axis and therefore end-shape of quartz grains is an expression of its initial shape. The elongation of quartz grains is considered as a criterion of provenance. The quartz from metamorphic rocks is generally more elongated than that from igneous rocks. The elongation of quartz grains therefore, has been considered as a criterion of provenance. A mean elongation ratio i.e. long axis/short axis of quartz grains from granites has been found to be 1.43, while that from schists is 1.75.

Another property of quartz is its extinction. It varies from sharp to wavy. Quartz that has been subjected to pressure or stresses, more commonly shows "strain shadows" or "undulatory extinction" in microsections, when observed under crossed nicols (Photo 9.2). Quartz from metamorphic rocks generally show wavy extinction and that from igneous rocks does not. It has, however, been observed that some quartz grains from granites or granitic rocks show wavy extinction or strain shadows, and therefore, extinction of quartz grains is not considered as a reliable guide for provenance.

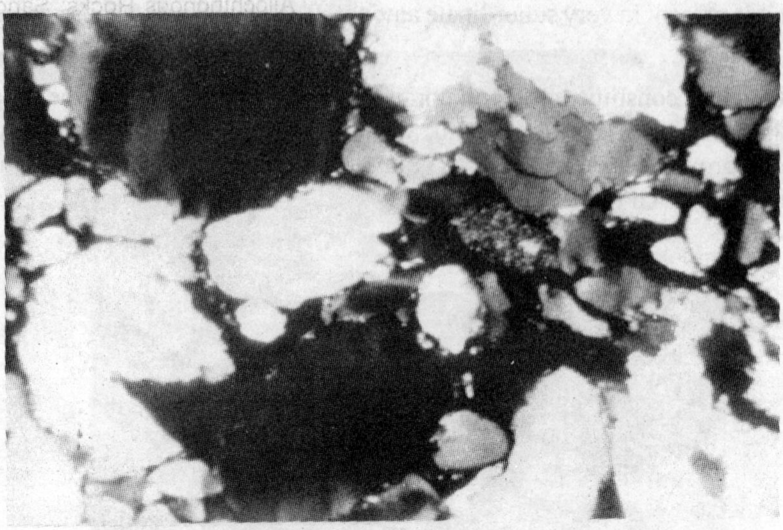

Photo 9.2. Quartz grain showing strain shadows and undulatory extinction, between crossed nicols, X60.

However, some of the features of quartz grains are considered to be more diagnostic of one provenance rather than the other. Thus, quartz derived from volcanic igneous rocks is typically monocrystalline, exhibits sharp, uniform extinction and is free from inclusions. Quartz of hydrothermal origin may be monocrystalline or coarsely polycrystalline but contains numerous fluid-filled inclusions. Polycrystalline quartz is considered to be derived from metamorphic rocks, that possess many crystals and are often elongated, with preferred crystallographic orientation. The extinction properties and polycrystallinity of quartz considered together help to separate grains of plutonic and low-grade and high-grade metamorphic origin.

Feldspar: Next to quartz but present in subordinate amount in sandstone is feldspar. Identification of feldspar and discrimination between different types in microsections, is rather a difficult task. Of the different feldspars, microcline, perthite and plagioclase feldspar can easily be distinguished. Detrital feldspar is either clear or partially to completely clouded with alteration products (Photo 9.3). Attempts have been made to relate a variety of feldspar to potential source rocks. Thus, oscillatory zoning in feldspar suggests volcanic or hypabyssal origin. Feldspar of acid volcanics is commonly sanidine, while that from plutonic rocks is microcline or orthoclase.

Feldspar is less stable than quartz, i.e. it is more susceptible to mechanical wear than quartz. Therefore, weathering products of plutonic rocks gradually get impoverished in feldspar, with a consequent enrichment of quartz. The feldspar is also chemically unstable; instability varying with the species. Lime-rich variety of feldspar is less stable than alkali feldspar, especially microcline. The alkali feldspars, i.e. Na- and K- feldspars are stable at low pressure-temperature conditions and in aqueous environment that generally prevail in sandstone.

Like quartz, feldspar is generally derived from the same crystalline rocks, namely; granites and granite gneisses. It is more commonly the derivate of the first sedimentary cycle. Apart from the contribution of feldspar from suitable source rock, feldspar content in sediments is also found to be controlled by the rate of erosion and climate. Humid climate in the source region brings about

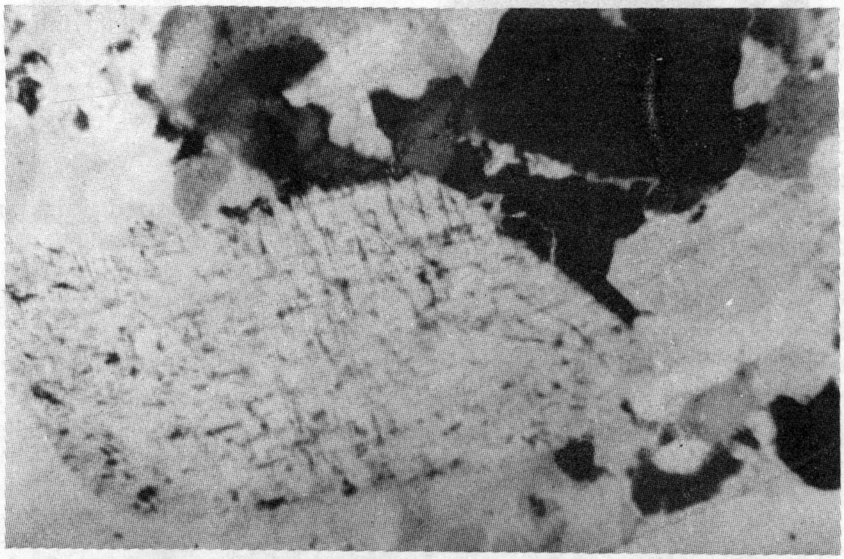

Photo 9.3. Altered feldspar grain in sandstone, X nicols X65.

rapid destruction of feldspar due to chemical weathering; while in arid climate, fresh feldspars survive mechanical weathering. The survival of feldspar in sediments, therefore, is a function of both, the intensity of decay process and duration of the action. Thus, with the high relief and rapid rate of erosion, feldspar escapes complete decomposition and it appears in sediments. However, with a low relief and retarded rate of erosion, feldspar gets destroyed. Though, the role of climate in feldspar decomposition appears to be secondary, under extreme arid or very cold climate, destruction of feldspar is retarded.

Secondary overgrowth on detrital felspars are also known but these form only a small fraction of the whole rock. Overgrowth on earlier feldspar nucleii is of pure soda feldspar, i.e. albite; and is in crystallographic continuity with the core. However, it generally shows a small difference in extinction angle, indicating that there is a small little difference in composition. Authigenic feldspar is always pure alkali feldspar.

Rock Fragments: Rock fragments, if present as clastic particles in sediments, are considered as important and significant; as they yield a definite information about source rock geology- provenance. These are the fragments of pre-existing rocks in sediments and should be described in some detail; so as to describe the nature and geological characteristics of the source region. Thus, if it is a metamorphic rock fragment, it can be inferred that source rock area is regionally metamorphosed and probably had attained upper amphibolite facies conditions or as the case may be.

Generally, coarse grained igneous and metamorphic rocks do not occur as detrital grains in medium grain-sized sediments/or in sandstones. Such sediments are primarily derived from the disintegration of plutonic rocks; whereas fragments of fine-grained rocks may appear in some lithic arenites. As the rock fragments are strongly size-dependent, they commonly occur in coarse-grained rocks, such as; conglomerates and breccias. As grain size decreases, identification becomes more and more difficult. At the same, as the rock fragment undergoes increased transportation, it is more likely to split into separate, individual minerals.

On the basis of textural characters, Dickinson (1970) has recognised the following types of rock particles; as:

1. volcanic rock fragments, with a texture of igneous aphanites;
2. clastic rock fragments, showing fragmental texture;
3. tectonite fragment with schistose or semi-schistose fabric;
4. microgranular rock fragments, having roughly equant, well-sized grains, and
5. carbonate rock fragments.

These categories of rock fragments can further be subdivided, according to further classification of rock type in each class.

Rock fragments generally have varied durability. Chert and felsitic rock fragments, due to higher resistance power to abrasion, tend to survive longer transportation (Photo 9.4), whereas rock fragments of other categories, as stated earlier, tend to become smaller and smaller in size and may even yield individual minerals, loosing its original identity.

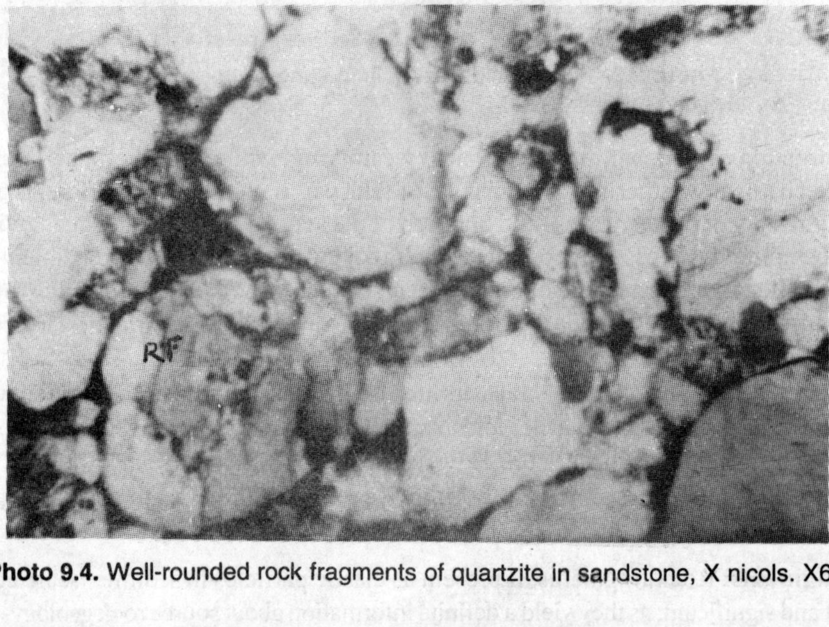

Photo 9.4. Well-rounded rock fragments of quartzite in sandstone, X nicols. X60.

Heavy Minerals: Among the different minerals of the source rock, those which survive longer, extensive transportation and therefore abrasion, are the 'heavy minerals'. These minor accessory minerals in sandstone are characterised by higher than average specific gravity, i.e. greater than that of bromoform - 2.85. These are chiefly silicates and oxides and many of which are much resistant to chemical weathering and mechanical abrasion. Heavy minerals are of two types, mainly opaque and non-opaque heavy minerals. Ilmenite and magnetite are two common opaque minerals; while common non-opaque heavy minerals are epidote, garnet, sphene, staurolite, rutile, tourmaline and zircon. These heavy minerals, as accessary grains, are present generally in concentrations less than 1.0 %.

Study of heavy minerals yields useful information about provenance-source rocks and of events that had taken place in the source region. Thus, certain heavy minerals are characteristics of some specific source rocks. Presence of garnet, epidote, staurolite, for example, indicates their derivation from metamorphic terrain; whereas, rutile, tourmaline, zircon indicate that these are derived from acid igneous rocks. Study and examination of heavy minerals should be carried out in sizes, as small as, 10 um. Commonly, sand-size fractions are examined, but heavy minerals, being finer in sizes, are also required to be studied from silt-size sediments. For the initial separation of heavy minerals, bromoform is frequently used. For further separation, other heavy liquids are tetrabromoethane—sp. gr. 2.95; methylene iodide—sp. gr. 3.30, and Clerici solution—sp. gr. 4.25.

Various mineral species that are encountered in clastic rocks also exist in the heavy mineral suite. However, number of mineral species and their relative amounts vary on different scales of observation. On the basis of mineralogy and mineralogic variations, heavy mineral provinces can be identified.

A sedimentary petrologic province consists of a group of sediments that constitute a natural unity by age, origin and distribution. Defining such provinces help in recognising trends in regional distribution and stratigraphic sequence. The petrologic provinces are gradational in both, i.e. in space and time; and are generally characterised by the presence of most abundant mineral species.

A mineral association, therefore, is an association of minerals by which a sediment or petrologic province is characterised. Granular variations are local concentrations of a particlular mineral species that has resulted due to differences in grain density, size or shape. Such variations are generally superimposed on mineral associations.

Heavy minerals can be dissolved during sandstone diagenesis through intrastratal solutions. This indicates that older sandstones tend to have a less diverse heavy mineral suite.

9.3 CEMENTATION AND DIAGENESIS

Diagenesis is divided into two broad stages; an early diagenesis for processes that take place from deposition into shallow burial realm and late diagenesis for those processes which affect the sediments at deep levels, and during and after uplift. The terms syn- or eodiagenesis, ana-or mesodiagenesis and epi-or telodiagenesis are commonly used for early and burial diagenesis and for diagenesis after uplift, respectively.

Significant physical processes of diagenesis are compaction and pressure solution and both are dependent mostly on depth of burial.

Generally, majority of chemical reactions take place on the earth's surface, within a very narrow range of temperature, yet not more than 40°C; whereas, burial temperatures range sometimes more than 200°C and transgressing in the realm of metamorphism. However, no single criterion can differentiate between 'sedimentary' processes from 'metamorphic' processes. Thus, e.g. $CaCO_3$ in sediments may recrystallize at near surface temperatures; montmorillonite may change to illite at about 100°C, while detrital or diagenetic quartz recrystallizes above 150 °C. Minerals so formed in a particular diagenetic environment remain stable in that environment, unless and until there are no further changes in the physico-chemical parameters.

In diagenesis, chemical processes bring about the precipitation of minerals, generally leading to the cementation of sediments, dissolution of unstable mineral species and replacement of one mineral by another. These processes usually take place in the medium of water, so that salinity, pH (acidity-alkalinity index), and Eh (redox potential) of water and rate of movement of water through sediments (dependent on porosity and permeability) play a critical role in diagenesis.

The early stage of diagenesis lasts for about 1000 to 1,00,000 yrs and affects sediments to depths of about 1 to 100 m. Pore fluids in sediments are related to depositional environment; e.g.

connate waters, if it is marine environment; while fresh water in the case of majority of continental sedimentary environments. These pore waters get modified due to breakdown of organic matter and bacterial activity. In marine sediments, early diagenesis takes place in oxidising pore water, that becomes reducing with depth, as oxygen is used in bacterial processes.

In case of deep burial, pore waters are further modified due to reaction with clay minerals, dissolution of unstable minerals, precipitation of authigenic minerals and mixing with waters from other sources. Processes of diagenesis of such nature generally require millions of years and affect sediments to depth of about 10,000 m, where temperatures at such depth range generally between 100-200 °C. With further increase in depth and therefore temperature, realm of metamorphism generally takes over.

Sandstone diagenesis is generally controlled by depositional environment, texture and composition of the sediments, migration of pore fluid, burial history, etc. Major changes that are involved in diagenesis are compaction and pressure solution, precipitation of calcite and silica cement, and also ferrugenous cement, authigenesis of clay minerals, overgrowths on detrital quartz grains, etc.

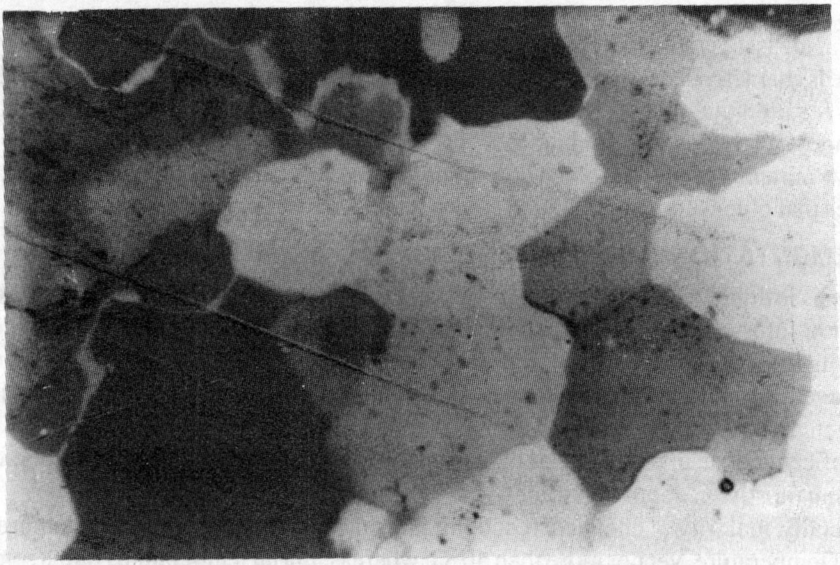

Photo 9.5. Compact sandstone showing a variety of grain contacts, X nicols X60.

The initial stages of burial of sediments involve compaction that results into dewatering and closer packing of grains. Further compaction through overburden results into local fracturing and/ or bending of weaker grains. At the same, the burial of sediments is associated with an increase in temperature. The change in temperature and pressure brings out the change in solubility of mineral species; e.g. with an increase in temperature, there is a decrease in solubility of calcite, resulting into precipitation of mineral from the pore fluid. Due to increase in pressure, the solution of grains at points of contact may also take place. Pressure solution at grain contacts is nominal, where the sediment is already cemented before deep burial, or if the matrix is present in the sediment, as in such case, load or overburden pressure is distributed and grain contact pressure is reduced. Once the sediment is cemented, the pressure solution may take place to produce irregular or sutured planes, which are termed as stylolites (Photo 9.6). Grain contact pressure solution and stylolites are the processes of significance, as these provide the material for cementation of the sediment. Also, pore fluids migrate in response to pressure gradient; mixing of different fluids may result in supersaturation, though neither of the fluid was initially saturated with respect to a particular mineral species. Thus, with the burial of sediment, precipitation of new mineral species and lithification of sediments take place.

Different types of cements are encountered in sandstone. However, there appears to be no particular sequence of cementation. With the change in temperature, chemistry of pore water, pH and other parameters, earlier cement is replaced by later one; and therefore, a sequence of replacement of cements in one sandstone may be reversed in another sandstone. However, discussion of cementation and diagenesis usually begins by noting the nature of previous cements.

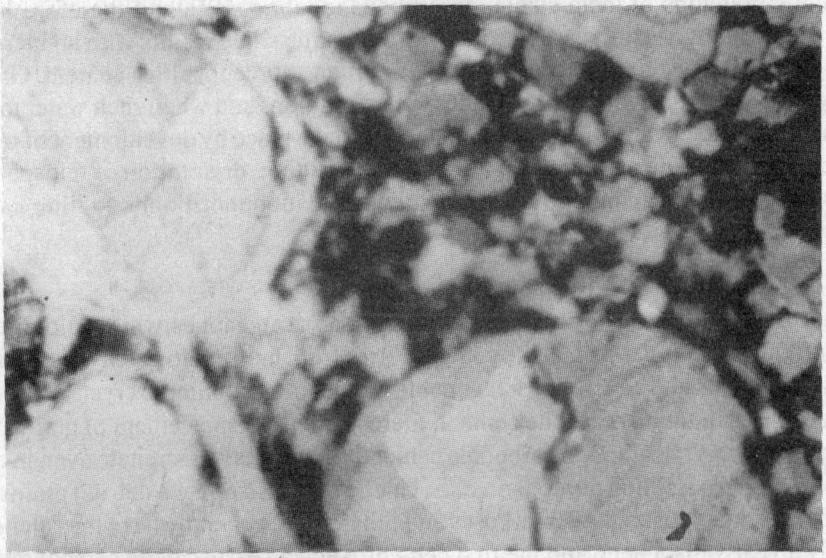

Photo 9.6. Quartz grain showing stylolite-pressure solution ettect in sandstone, X nicols, X65.

Silica Cement

Diagenetic silica in sandstone appears in various morphologic and crystalline forms. Cristobalite is one of the crystalline forms, that occurs as fibrous rim cement around detrital grains in volcanic sandstones. This type of cement has generally been identified in sandstones that contain volcanic detrital components. Quartz that is formed during diagenesis may occur as microcrystalline, subhedral grains of chert or as a coarsely crystalline mosaic of subequant grains or as elongate fibres of chalcedony or as secondary euhedral overgrowth on detrital quartz grains or even as microquartz replacement by carbonate minerals; (either as fossils or as matrix) of the diffferent morphologies of quartz. It has been observed that macrocyrstalline morphology seems to be most stable.

The most common type of silica cement is quartz overgrowths on detrital quartz grains. It is precipitated around the quartz grain, and is in optical continuity of the host. Such syntaxial overgrowths generally yield euhedral crystal boundaries to the host grain. In microsection, the boundary between overgrowth and host grain is marked by impurities, which may be iron oxide, clay, fine dust particles or water bubbles (Photo 9.1).

Origin of silica for such cementation is generally attributed to pressure solution. It is pore fluid that becomes enriched in silica and after attaining the stage of supersaturation, it is deposited as silica cement in the form of overgrowths on detrital quartz grains. Quartz overgrowths without pressure solution may indicate the possible sources of silica cement, as; dissolution of silica dust, silicate or biogenic silica and groundwater.

It might be the dissolution of feldspars, pyroxenes and amphiboles that would provide silica or even the mineral transformations, such as; montmorillonite to illite and feldspar to kaolinite provide silica for cementation. In marine sediments, pore waters often contain significant amount of silica that is derived due to dissolution of diatoms, radiolarians and sponge spicules. These have the siliceous skeletons of meta-stable, amorphous, opaline silica and therefore having higher solubility than quartz. An increase in solubility of opaline silica results with an increasing depth, because of which biogenic silica could be an important source for silica cement. Groundwater is generally superstaturated with quartz. Under such conditions and when such water moves through quartz sand in sufficient quantities, cementation may take place by development of overgrowths.

Sources of silica cement, therefore, are pressure solution, dissolution of feldspars, pyroxenes and amphiboles and even mineral transformations like montmorillonite to illite and feldspar to kaolinite, etc.

Calcium Carbonate Cement

Of the carbonate cements, calcite is the most common in sandstones;while dolomite and siderite are rather of local significance. Calcite cement varies from uniform to patchy distribution and also to local segregations and concretions. It generally appears as poikitopic crystals and drusy calcite spars. Due to precipitation of calcite cement, there is often a displacement of grains, that gives an appearance of grains 'floating' in carbonate cement. Calcite may precipitate even in cracks within the grains. Apart from filling pores and voids, calcite may also replace detrital grains.

Sandstones, like quartz arenites, arkoses and litharenites, generally have the calcite cement.It is frequently an early diagenetic and the first cement. Such an early calcite cement rather prohibits later quartz overgrowth formation and also feldspar alteration, but at the same, it can result in total loss of porosity and permeability in sand or in sandstone. In other sandstones, calcite cementation may post date quartz overgrowths, rather due to late increase in the activity of the carbonate ion.In a very shallow subsurface region, $CaCO_3$ precipitation may happen through evaporation of vadose or near surface phreatic groudwater; whereas at depths carbonate precipitation can be brought about by an increase in pH and/or temperature.

Chemical relationships between calcite and quartz, as the replacement of one mineral by the other suggests that pH and temperature are the most common controlling parameters. Solubility of silica is generally unaffected by pH value less than that about 9.0; whereas calcite is soluble in acidic solutions. It is, therefore, evident that changes in hydrogen ion concentration causes solubility of one phase with the simultaneous precipitation of the other.

However, with an increase in temperature, there is an increase in silica solubility. Groundwater or pore water containing calcite, on heating, may generate a gas phase, i.e. carbon dioxide and its escape causes solubility of calcite to decrease with increasing temperature. Thus, as temperature increases, the net result is a decreased solubility of $CaCO_3$.

Calcium carbonate cement is rather uncommon in Holocene silicate sands. It is either aragonite or calcite. It is uncommon as silicate sands are deposited rather in relatively agitated water and when any soft or cleavable carbonate gets deposited on grain surfaces, it is removed very easily. In ancient sandstones, however, dominant carbonate cement is low magnesian calcite. Studies have indicated that with depth, a transition may take place from low magnesian calcite through ferromagnesian calcite to dolomite.

Hematite Cement

In some sandstones, red pigment matter is present as a primary pore filling that is precipitated through solution, as microcrystalline iron oxide, deposited alongwith the clastic grains (Photo 9.7) in continental environments. Hematite generally occurs as a very thin coating around detrital grains. It is chiefly amorphous or consists of micron-size crystals. Main sources of hematite cement or red pigment are Fe-bearing minerals of igneous and metamorphic rocks; such as; amphiboles, pyroxenes, olivine, mica-mainly biotite, chlorite, ilmenite and magnetite.

According to one theory, hematite pigment in red beds is considered to be of detrital origin. Amorphous iron is considered to be formed in moist, tropical, lateritic weathering conditions in upland areas, which is later transported and deposited alongwith the sediments and then converted to hematite.

However, another theory proposes purely diagenetic mechanism for the origin of hematite cement. Accordingly, iron is supplied by intrastratal solution of Fe-bearing silicate minerals. If it is the oxidising diagenetic environment, iron is reprecipitated as hematite or a goethitic oxide phase as an early phase that later converts to hematite. Development or formation of hematite generally takes place above the water table and below, if the groundwater is alkaline and oxidising. In reducing conditions, however, iron is present in more soluble ferrous state and if it gets incorporated in clays, it imparts green colour to the sediments.

Eventhough, an intrastratal-diagenetic origin of hematite is considered as an explanation for its presence in red beds; in matrix-rich sandstones, however, its detrital origin seems to be more viable.

The preceding paragraphs suggest that in analysing diagenetic changes in sandstones, controlling or significant parameters are temperature gradient, depth of burial, original lithological characters of sediments, composition of interstitial fluid, and time. Temperature gradient and depth of burial both change with time, whereas depth of burial normally does not increase, unless the uplift of the region takes place. These factors indicate the complexities involved in interpretation of diagenetic changes in sandstone and all the changes recorded as diagenetic are to be explained as stages of diagenesis, which have gradational boundaries.

9.4 CONGLOMERATES AND BRECCIAS

Some rocks such as tillite or indurated boulder clay, containing less than 10% of gravel-sized fragments, are designated as 'conglomerates'. The term conglomerate is generally applied to indurated gravels. The term 'breccia' is used as the consolidated equivalent of rubble; where rubble is an accumulation of angular fragments, which are generally coarser than sand. The term breccia is also applied to non-sedimentary rocks, such as; fault breccia or volcanic breccia.

Generally, the term conglomerate is used where majority of the fragments are subangular and better rounded (Photo 9.8), whereas the term breccia is used only to those rocks in which the fragments are angular (Photo 9.9).

Composition of conglomerate or breccia can be made by counting and classifying the fragments on the basis of their genesis. Thus, the composition of the fragments can be represented by grouping the fragments into extrusive igneous, plutonic igneous, sedimentary and metamorphic classes. These are then plotted in a triangular diagram, generally with sedimentary and metamorphic fragments

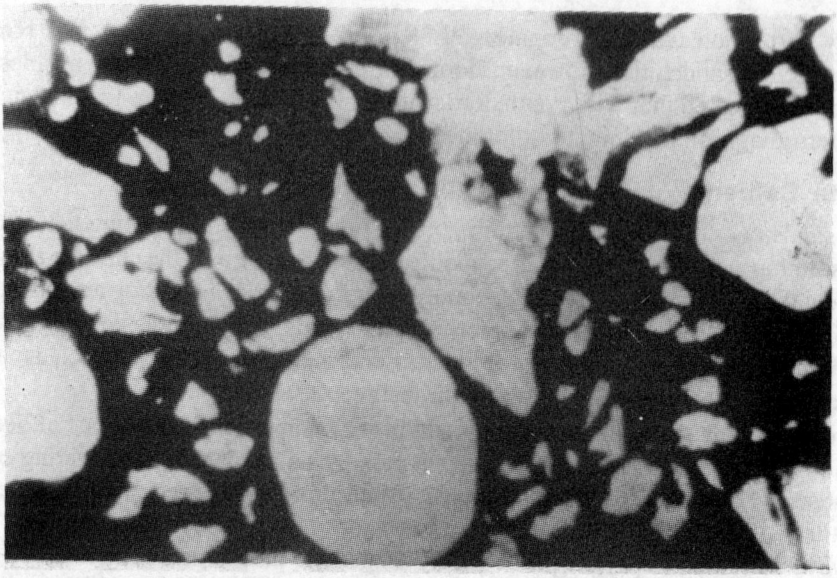

Photo 9.7. Sandstone with hematite cement, polarised light, X65.

Photo 9.8. Conglomerate, showing majority of rounded detrital grains.

considered together. However, a distinction is generally made between supracrustal and plutonic provenance. Such a distinction to some extent leads to a measure of the extent of uplift and depth of erosion in the source area, where both are functions of tectonics.

The composition of conglomerate or breccia, however, is not an exact expression of the kind and abundance of rock types in the source region; because of the varying block-forming capabilities of different rock-types, their varying resistance to abrasion and other processes of disintegration.On the basis of the composition and characteristics of the detrital constituents, these coarse clastic rocks can be classified, as; conglomerate with subrounded to well-rounded clasts and breccias with angular and subangular clasts.

Photo 9.9. Breccia, showing majority of angular detrital grains.

In terms of composition, these are identified as polymictic conglomerate (or breccias), which contain a variety of pebbles and cobbles; whereas, mono-or oligomictic conglomerate or breccias comprise pebbles of only one type.On the basis of the types of clasts; Intra-formational and Extraformational conglomerates have been identified. The Intraformational clasts are formed within the area of sedimentation and are chiefly lumps of mud that are derived due to reworking and recycling of previously deposited mud; whereas in case of Extraformational clasts, they are derived from outside the area of sedimentation and consist of any type of rock, even at times very unstable varieties, if the transport is short.

On the basis of the fabric, two types have been identified, as clast-supported and matrix - supported conglomerates. In sediment fabric, a variety of grain contacts occur; such as; point contacts, concavo-convex contacts, sutured contacts, etc. (Fig. 9.2, Photo 9.9) In some cases, where there is a lot of matrix, grain contacts may not result, but rather 'float' in the matrix. This is referred to as

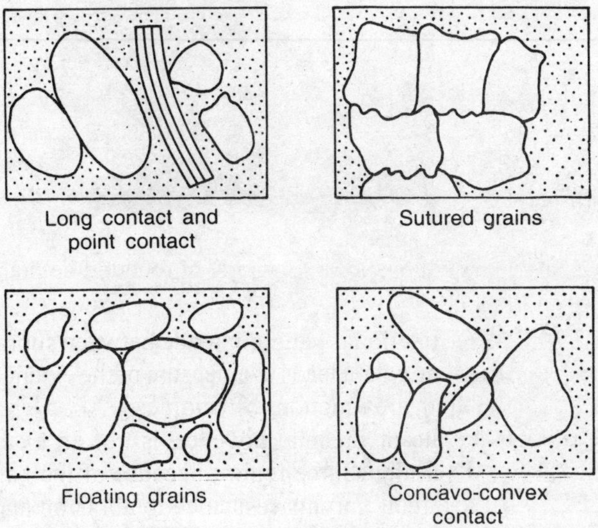

Fig. 9.2. Nature of grain contacts.

matrix-supported fabric. In case of the conglomerates of river channel, stream flood or of more aqueous flows, matrix is little, pebbles are in contact with each other, resulting into clast-supported fabric.

Additional, particular types of breccias have also been identified, as slump breccias and solution breccias.

When sediments are deposited on a very gentle to significant slopes, downslope movement results on a large or small-scale, that can be termed as slumping. It involves individual beds or strata. Due to such a downslope movement, brecciation of beds or strata takes place, involving practically the entire bed. After consolidation at the downslope, it gives rise to slump breccia.

Another type is solution breccia, where dissolution of subsurface evaporite beds takes place, causing the collapse and brecciation of overlying strata.

Thus, a variety of conglomerates and breccias can occur, which are classified on the basis of composition, nature, type and fabric of the constituent clasts.

10
CHAPTER

Autochthonous Rocks: Carbonate Rocks

10.1 INTRODUCTION

There are two broader divisions of sedimentary rocks in which they are classified; as Extrabasinal or allochthonous sedimentary rocks, originating outside the basin of deposition and Intrabasinal or autochthonous sedimentary rocks, originating within the basin of deposition. Chemical rocks are those which are classified as autochthonous or intrabasinal rocks.

Chemical rocks are further divided into two types, as:

(*i*) organic type and
(*ii*) inorganic type.

Organic type of chemical rocks include mainly carbonates, while inorganic type includes evaporites. However, carbonate rocks are also found to originate by inorganic processes, i.e. by direct precipitation and hence there is no clear cut defined boundary between these two types.

The following are the major types of chemical rocks:

I Carbonates: Limestones and dolomites
II Evaporites: Anhydrite/Gypsum Halite/Rock salt; potash salts, etc.
III Siliceous Rocks: Chert, radiolarite, novaculite.
IV Carbonaceous Rocks: Humic Group : Coal series. sapropelitic Group: Oil shales and Cannel coals.
V Sedimentary Ironstones
VI Phosphates

Of these, carbonate rocks are found to be volumetrically most abundant. Compositionally, these are predominantly made up of calcium carbonate, $CaCO_3$ and the dolomite, composed of the mineral dolomite—$Ca,Mg(CO_3)_2$. In order to distinguish between a mineral and a rock, the term 'dolostone' is used for the rock.

Biological and biochemical processes are responsible in the formation of carbonate rocks. However, carbonate rocks are formed also by direct inorganic precipitation from sea water and also by diagenesis. Economic importance of limestones chiefly lies in the fact that they are important aquifers and present hydrocarbon reservoirs, sometimes due to high porosity.

Carbonate rocks are, thus, complex in their genesis and also in diagenesis. They have a wide spectrum of particle sizes, that range from whole shells to lime mud of diverse origin. The carbonate sediments are initially deposited with high primary porosity. Carbonate minerals that are formed are also chemically unstable. Combination of high primary porosity and permeability, alongwith the unstable carbonate minerals leads to complicated diagenetic processes; giving rise to a variety of complex petrophysical characters.

10.2 CARBONATE MINERALS

Calcium carbonate ($CaCO_3$) is the dominant constitutent of modern and ancient limestones, and it occurs as calcite (hexagonal) and aragonite (orthorhombic). Calcite forms isomorphous series with magnesite ($MgCO_3$). However, two types of calcite are recognised, depending on the magnesium content, as low magnesium calcite with less than 4 mole % $MgCO_3$ and high magnesium calcite with greater than 4 mole % $MgCO_3$, generally ranging between 11 to 20 mole % $MgCO_3$. Yet; distinction between high and low magnesium calcite is arbitrarily made at 10 mole %. Modern carbonate sediments more commonly contain aragonite and high magnesium calcite; while ancient limestones are largely made up of low magnesium calcite. Since aragonite is metastable and high magnesium calcite loses its Mg through time, all carbonate sediments are gradually converted into low magnesium calcite.

Another important carbonate mineral is dolomite having chemical composition, as; $Ca,Mg(CO_3)_2$. Isomorphous substitution of some Mg for Fe is found in the mineral ferroan dolomite or Ankerite $Ca(Mg, Fe)(CO_3)_2$.

Some common carbonate minerals, alongwith the composition and occurrence of each have been presented in Table 10.1

Table 10.1: Common Carbonate Minerals

Name	*Composition*	*Occurrence*
Aragonite	$CaCO_3$	Present in certain carbonate skeletons, unstable and reverts to stable polymorph calcite
Calcite	$CaCO_3$	Present in certain carbonate skeletons, as mud (micrite) and sparrite (cement)
Magnesite	$MgCO_3$	Present in minor amounts within the lattices of skeletal aragonite and calcite.
Dolomite	$Ca\,Mg(CO_3)_2$	Largely as a crystalline diagenetic rock, also pene-contemporaneously associated with evaporites.
Ankerite	$Ca(Mg\,Fe)\,(CO_3)_2$	A minor variety of dolomite. i.e. Ferroan dolomite
Siderite	$FeCO_3$	As concretions and ooliths.

10.3 NOMENCLATURE AND CLASSIFICAION

Various schemes for classifying carbonate rocks have been suggested. For the purpose, the following main components of carbonate rocks are taken into consideration:

(*i*) Particles; meaning sand size grains, and include skeletal and non skeletal grains; which are identified as follows;

- Grains
  - Detrital grains
    - Lithoclasts
    - Intraclasts
  - Skeletal grains
  - Peloids (including faecal pellets)
  - Lumps
    - Composite grains
    - Algal lumps
  - Coated grains
    - Ooliths
    - Pisoliths
    - Algally encrusted grains

(*ii*) Micrite is another major component of carbonate rocks. It is lithified, mechanically deposited lime mud. It occurs either as matrix among sand size particles in a limestone or as only particles in fine-grained limestone. In micrite, particle size is usually less than 4 μ, but commonly lies between 1 and 3 μ. Electron microscope studies indicate that it is not homogeneous but it has areas of finer or coarser crystals. The matrix in limestone is thus composed of either micrite or clay.

(*iii*) Another component is cement. It is normally chemically precipitated pore-filling and consists of sparry calcite. Micrite is generally susceptible to diagenetic alteration, being replaced by coarser mozaic of microspar through the process that is termed as aggrading neomorphism.

Folk (1962) has classified these components broadly into two groups, as:

(*i*) Allochemical constituents, or simply as allochems that include discrete carbonate grain/particles or aggregates; and

(*ii*) Orthochemical constituents, or simply as orthochems, that include micrite and sparry calcite cement.

Classification systems for limestone that are commonly used, have a different emphasis.

1. A simple scheme of classification divides limestone on the basis of grain size into:

calcirudite, where most grains are more than: 2mm. in size;
calcarenite, where most grains lie between 2mm. and 64 μm., and
calcilutite, where most grains are less than 64 μm.

Another scheme of classification, suggested by Folk (1959, 1962) is based on allochemical and orthochemical constituents. In this classification, four kinds of allochemical constituents; i.e. kinds of particles are recognised, as:

a. Intraclasts, including grapestones.
b. Ooids,
c. Whole skeletons or fragments of fossils,
d. Pellets.

The interstices among these four kinds of particles are filled with either sparry calcite cement or micrite. On the basis of the combinations of four kinds of particles with micrite or cement, eight different kinds of limestones have been recognised. In addition, micrite may lack the presence of sand-size particle and therefore, it is a ninth kind of limestone and tenth kind is a Reef rock.

The classification of limestone, as suggested by Folk is presented in Table 10.2:

Table 10.2

Allochemical Rocks			Orthochemical Rock	Reef Rock
Sparry Calcite Cement		*Micrite*		
	I	*II*	*III*	*IV*
Intraclast	Intrasparrite	Intramicrite	Micrite	Biolithite.
Ooids	Oosparrite	Oomicrite		
Fossils	Biosparrite	Biomicrite		
Pellets	Pelsparrite	Pelmicrite		

Modified version of this scheme of classification, suggested by Folk, is based as relative proportions of lime mud, i.e. micritic matrix and spar cement. The same is presented in Table 10.3

Table 10.3: Classification of Limestone

	Over2/3 lime mud matrix				*Subequal spar and lime mud*	*Over 2/3 spar cement*		
Allochem %	*0-1%*	*1-10%*	*10-50%*	*Over 50 %*		*Poor sorting*	*Sorting good*	*Rounded & abraded*
Representative Rock term	Micrite or Dismicrite	Fossiliferous micrite	Sparry biomicrite	Packed biomicrite	Poorly washed biosparite	Unsorted biosparite	Sorted biosparite	Rounded biosparite
		---------- Biomicrite	---------	---------		Biosparite ---------		

Another scheme of classification that is commonly referred to is that suggested by Dunham (1962), which is based on the texture of limestone. Accordingly, grainstone is the one with carbonate grains without matrix; such as a bio-or oosparite; packstone is the one with grains in contact with matrix, such as a biomicrite; and wackestone that has coarse grains floating in a matrix, similar again to a biomicrite and mudstone is the one with predominant micrite with few grains present.

Several additional terms have been introduced by Embry and Klovan (1972), in the scheme of classification suggested by Dunham.

As has been stated, limestone classification is based on three major components; viz., carbonate sand sized particles, micrite and sparry calcite. Of these, a variety of carbonate sand-sized particles occur in limestone. These are described in the following paragraphs.

Skeletal Particles: Skeletons of invertebrates and calcareous algae may produce carbonate sand-sized particles. Presence of such skeletal particles is a reflection of the distribution of carbonate secreting invertebrates through time and space. The specific environment of formation and environmental factors, such as; depth, temperature, salinity, substrate and turbulence determine the type of skeletal grains that are encountered in limestone. Thus, e.g. during Late Palaeozoic, crinoid thickets produced abundant crinoidal limestone. Green calcareous algae-Halimida could produce grains that accumulate as carbonate sediment in modern seas on shallow water banks and lagoons.

Skeletons of such organisms may break into smaller grains due to transport by waves and currents. Organisms also eat sediments for the organic content and that causes the disintegration of internal structure of skeletons. Decomposition of the muscle fibres and ligaments in bivalves separate the shell, further giving rise to fragments (Photo 10.1).

Principal carbonate secreting organisms and their age range are as follows:

Algae	Late Palaeozoic	Mostly calcite
	Mesozoic	Calcite and aragonite
	Cenozoic	Aragonite and calcite.
Foraminifera	Late Palaeozoic	Calcite
	In Late Mesozoic but mostly in Cenozoic	Calcite >> Aragonite
Brachiopoda	Abundant in Palaeozoic	Calcite
	Less common in Mesozoic	Calcite
Trilobites	Upto Middle Palaeozoic	Calcite

These various microorganisms have played a significant role in the formation of carbonate deposition by secreting either aragonite or calcite, as indicated above.

Ooids: These are spherical-subspherical sand-sized $CaCO_3$ particles, having one or more regular concentric rings of $CaCO_3$, surrounding a nucleus of another particle. Such a nucleus is either often a carbonate particle or quartz grain (Photo 10.2). In modern environments, the oolitic coatings are found to be of aragonite. Sediment composed of ooids is referred to as an oolite. The term, ooid has been restricted to grains, which are less than 2 mm. in diameter. Majority of ooids range from 0.2 mm. to 0.5 mm. in diameter.

Photo 10.1. Fossiliferous limestone.

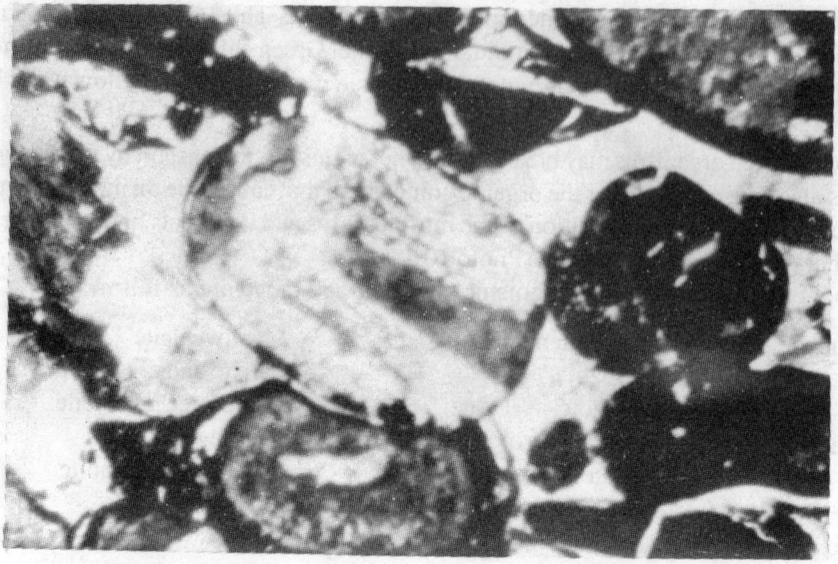

Photo 10.2. Limestone showing oolitic grains, in polarised light, X55.

At the time of formation, individual crystals of aragonite get oriented and are tangential to the surface of the grain. In ancient ooliths, preservation of microstructures may be seen, indicating that they were originally formed as calcite. Ooliths that are found today are formed where only strong bottom currents operate, i.e. in agitated water conditions. Depths of water, where ooids occur, are usually less than 5 m; but they may even reach upto 10-15 m., thereby indicating the areas of formation as tidal bars or tidal deltas, or form shoals close to the edge of the platform. Formation of ooids has also been reported from lagoons, lakes and rivers and on tidal flats, in fresh and hypersaline water. The presence of large scale cross-bedding in ancient carbonate deposits is an evidence for current transport.

In the depositional sites, such as tidal bars and tidal inlets, the formation of ooliths is obvious. In such an environmental setting, grain can be moved periodically by remaining in the same depositional site for a sufficiently long period of time to develop a thick oolitic coating. Such a coating on a particle is evidence for current transport, in an environment, where the grain is periodically buried and exhumed from the sediments. The thickness of coating on a grain is an indication of the length of time for which the grain was subjected to the process.

Pisolith: These are large ooliths, and are considered mainly as diagenetic features, originating in the vadose zone. After the pisoliths are formed, some of them get separated from the sediment or rock and are redeposited as particles in new sediments.

Peloids: Peloids are the most common carbonate particles in modern day carbonate sediments and also in ancient limestones. These are aggregates of microcrystalline calcium carbonate without having any internal structure. Majority of peloids are ellipsoidal to roughly spherical in shape (Photo 10.3) and may contain small fragments of skeletal material. These very much resemble to small rounded aggregates of carbonate mud. They may result from several processes and therefore, their origin is uncertain. Hence, the non-genetic term-peloid has been suggested by McKee and Gutschick

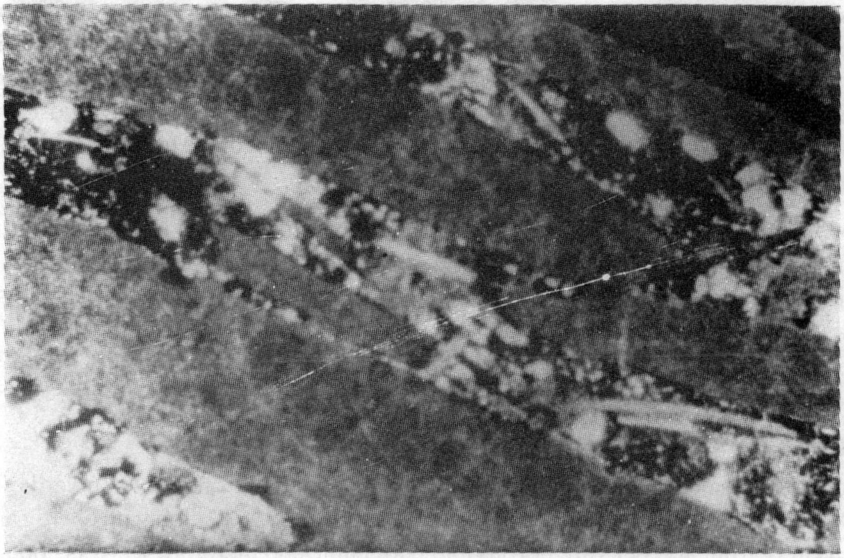

Photo 10.3. Limestone showing pellets, ellipsoidal or roughly spherical grains, in polarised light, X55.

(1969). Many of the peloids are fecal pellets, containing organic matter and assorted detritus, normally by carbonate mud ingesting organisms. Studies of modern carbonate sediments show that polychaete worms, the Gastropod Batillaria minima and some Crustaceans contribute to fecal pellets. These fecal pellets are more commonly poorly consolidated friable aggregates. In ancient carbonate deposits, many of the peloids show vague and diffused outline, which indicate that particle has been worked over number of times by organisms. Lithification of pellets can also be the cause for diffused outline, however, for which no satisfactory explanation has been sought.

It has also been observed that similar-looking particles, termed as pelletoids, can form due to micritisation of other particles, such as; skeletal grains or ooliths. In such particles, original internal strcture in most cases is preserved to indicate its identity; whereas in advance stage of micritisation, original particle may become uniform and homogeneous, having an appearance of micro-crystalline carbonate grain.

Aggregated Lumps and Grapestones: These are the composite particles, consisting of two or more peloids or ooliths, which are bound together by organically rich, tan to brown-coloured aragonite. Their formation generally takes place in shallow water carbonate platform environment. Such composite particles are termed as lumps. The term grapestone is used where there are rounded protruding grains, resembling a bunch of grapes. Their binding mechanism results due to growth activities in the substrate of organisms, such as foraminifera or microscopic worms. The presence of blue-green algae may also act as the binding material.

Algal Accretionary Grains: Carbonate sand grains may serve as growth surface for primitive blue green algae and related organisms. The algae then acts as a sticky surface, in which fine carbonate sediment particles get trapped. In such sediments, algal structure is seldom preserved and only the carbonate sediments surround the original particle, as irregular, discontinuous layers. Such grains are called as oncolites. Oncolites of larger size, maximum upto 8 mm. have also been reported and are termed as algal biscuits. These are similar in origin as that of algal stromatolites.

Intraclasts: The intraclasts are fragments of partially lithified carbonate sediments, which have been derived due to erosion of the basin floor or of the adjacent tidal flats, and vary in size and shape. The process of lithification and disruption of the particle from its original setting and redeposition essentially results contemporaneously with the sedimentation of the host stratigraphic unit, in which it is found. The term, 'intra', indicates the formation of the particles within the general area of deposition of the host sediments.

Lithoclasts: Lithoclast is a rock fragment that is derived from outside the basin of carbonate deposition. It is derived due to erosion and transported to the basin. Generally, lithoclasts are not very common in carbonate rocks. If the carbonate platform terminates landward against a cliff of older rocks, its erosional products may contribute to lithoclasts in new sediments.

To overcome the difficulty in distinction between intra- and lithoclasts, the term 'limeclast' is used.

10.4 ORIGIN OF LIMESTONE

The carbonate rocks occur in every geological period from the Proterozoic onwards. They have important economic significance due to their composition and association with energy and mineral deposits. Many carbonate rocks are porous and permeable and therefore serve as good reservoirs for groundwater and petroleum. It has been observed that nearly 50% of the hydrocarbon deposits of the world have been hosted by the carbonate fields.

Biological and biochemical processes play major role in the formation of carbonate sediments. After the deposition, chemical and physical processes of diagenesis considerably modify the carbonate sediments.

Limestone and dolostones constitute 10-15 % of the sedimentary rocks, with limestones being considerably abundant, compared to dolostone in most stratigraphic sections. In the Precambrian period, however, the limestones are often dolomitic and may contain algal stromatolites, produced largely by the blue-green algae. Carbonate rocks are normally free from impurities, which are generally less than 5 % of an average limestone and consist of fine grained quartz and clay minerals.

With few exceptions, most of the carbonate rocks observed on the continents have been deposited on shallow water platforms. In the geologic past, shallow seas periodically covered vast continental areas, thereby resulting into the deposition of carbonate sediments, mainly limestone, over thousands of square kilometers. Organisms with carbonate skeletons occur throughout the world's oceans resulting in the development of carbonate sediments anywhere, in oceanic environment. However, the most important factors, such as; temperature, salinity, water depth and input of siliciclastic material control carbonate deposition. Many carbonate skeletal organisms, such as; reef-building corals and calcareous green algae require warm waters to flourish. Majority of carbonate sediments, therefore, occur in the tropical and subtropical belts; i.e. nearly 30° north and south of the equator. Most limestones of the Phanerozoic period have formed in low latitudes. However, carbonate sands do occur in higher latitudinal areas.

Most of the carbonate skeletal organisms are affected by salinity and water depth and therefore preferentially occur in the shallow agitated part of the photic zone in sea water of normal salinity. Carbonate oozes are generally present in deep ocean floor, which are principally composed of pelagic organisms, that live in the photic zone. High rates of carbonate dissolution at depths of several kilometers result in few carbonate sediments below this depth.

The abundance of carbonate sediments has been recorded within the intertidal and supratidal environment. This is because a shallow water shelf or bank fill upto sea-level, if sedimentation is rapid and subsidence is relatively slow. Under such conditions, regressive carbonate cycle is produced from marine to intertidal, resulting into supratidal deposition. If carbonate platform rises from oceanic depth, supratidal sedimentation may result with a break or diastem that ends with next transgression. When a carbonate platform is attached to the continents, the supratidal carbonate sedimentation may be followed by continental sedimentation, prograding across the earlier tidal flat.

Deposition of calcium carbonate sediments is due to calcareous shell organisms, which are essentially marine and proliferate in areas of abundant light, constant salinity and clear warm water. For the purpose, therefore, water has to be very shallow and away from large rivers, which otherwise lower the salinity. Such a setting can be observed in shallow water shelf or bank, within the intertidal and supratidal enviroments, and also in platform shelf environments.

Most platform limestones are also shelf limestones, where the platform is shallow water marine extension of the continental block. Platform limestones may extend into the continental block as epeiric sea limestone, which generally cannot be distinguished from platform shelf limestones due to similarities in sedimentologic or petrologic characters.

The rate and type of carbonate sedimentation considerably vary across a carbonate shelf or platform. Physicochemical fixation and locus of skeletal production are generally concentrated along the seaward margin. Such a margin is favoured due to greater water turbulence and abundance of nutrients upwelling from deeper waters. Growth of reefs, therefore, commonly takes place along such margins. In the absence of reefs, ooids shoals are common, which on lithification become oosparrite. Carbonate production in areas landward of the reef or ooid shoal in the subtidal zone may also take place but due to agitated environment, preservation is poor and some carbonate sediment is transported landward on to the tidal flats, tidal channels and shore line complexes. In near shore environments, fine grained carbonate sediments like micrite, biomicrite and pelmicrite may accumulate, while further landward and above normal high tide or in the supratidal areas, formation of dolomite takes place.

10.5 LIMESTONE DIAGENESIS

The term diagenesis includes all the processes that operate after the initial deposition of carbonate sediments but before elevated temperature and pressure that create minerals and structures, considered to be in the realm of metamorphism. Generally, physical, chemical and organic processes operate immediately on carbonate sediments, following deposition, resulting into changing mineralogy and giving rise to sedimentary structures and textures, cementation and replacement with calcite, dissolution of original and diagenetic minerals and replacement with dolomite are the major processes that operate to bring about limestone diagenesis.

Diagenesis of carbonate rocks therefore, is accomplished by one or more of the three following operations, namely; dissolution, cementation and replacement.

The process of dissolution gives rise to void space within the rock by removal of earlier carbonate minerals in solution (Photo 10.4).

Cementation is generally considered as the growth of new crystals into void spaces, irrespective of the origin of void space.

Replacement essentially includes simultaneous dissolution of existing mineral and precipitation of new mineral.

The processes stated above are merely the generalisation by which diagenesis of the limestone takes place. It is possible that one or more processes operate simultaneously after the carbonate deposition.

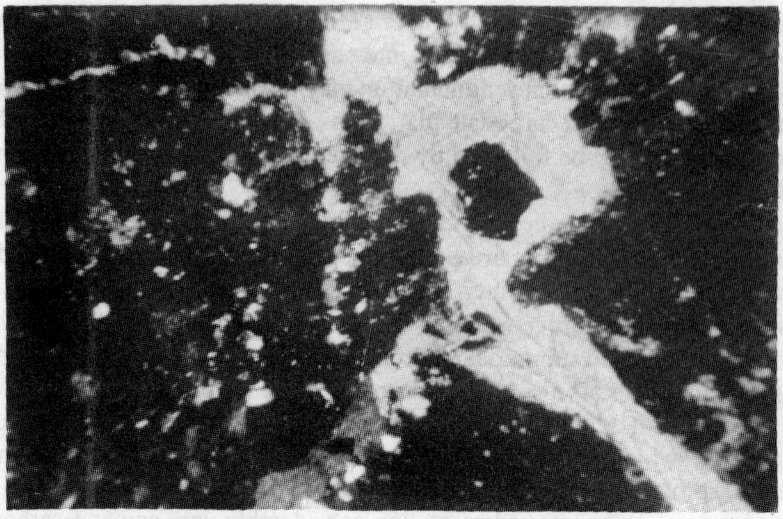

Photo 10.4. Reprecipitated calcite in vein, showing dissolution, cementation in polarised light, X60.

Dissolution: It has been observed that magnesium calcite and aragonite are stable mineral phases in surface sea water but are unstable in other environments. Similarly, in fresh water i.e. in the vadose or phreatic zone on the land or in most lakes, pure calcite is the stable mineral phase. In the deep sea, water is undersaturated with respect to all carbonate phases, where planktonic carbonate particles settle resulting into progressive dissolution of aragonite more rapidly than calcite.

Dissolution may be selective and aragonite from the skeletal part may only be removed or it may be non-selective where water movement takes place along joints, thereby enlarging the joints giving rise to enlarged cavity locally. In all cases of dissolution, it is presumed that pore water is not shallow sea water and at the time of dissolution, it is undersaturated with respect to the mineral phase that is removed in solution. It can also be presumed that due to relative low solubility of calcium carbonate in most natural waters, vast amount of pore volume of fluid had passed through the rock.

Dissolution of the interior of ooliths i.e nucleus and many of the precipitated coated rings is also a common observation. This indicates that material removed was either more soluble than aragonite at the time of dissolution or it was much porous.

Cementation: Cementation of carbonate rocks can occur very early in the history of sediments, on the shallow sea floor, within few 1000 years after formation, as the carbonate sediment is exposed to meteoric water by lowering of sea level. It is commonly accomplished by growth of new crystals into pore space from carbonate rock surfaces. The new crystals are equant, acicular or botryoidal. Their texture and growth appear to be controlled by degree of supersaturation of the pore water, rate of precipitation of calcium carbonate and also mineralogy of the cement. It is observed that under rapid fluid flow conditions, aragonite precipitation is favoured, whereas with the low rate of fluid flow, calcite is formed. In general, aragonite and sometimes high magnesium calcite can form as cementing crystals from sea water or a similar solution, while calcite crystals can grow as cement from fresh water solution. Principal environments of carbonate cements have been shown in Fig. 10.1.

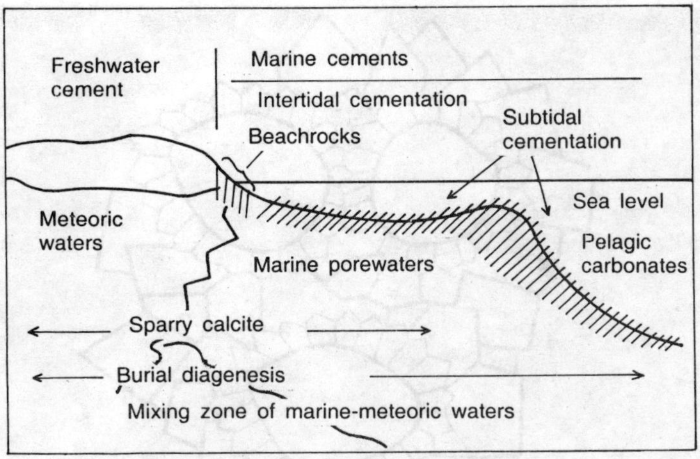

Fig. 10.1. Different environments of carbonate cementation.

Beach rock within the intertidal zone is found to be cemented by needle like crystals of aragonite that have grown from the grain surfaces into pore space between the grains. Aragonite grows from sea water, while it moves through the beach rock. Evaporation of sea water in contact with air within the pore space seems to be responsible for precipitation of the cement.

Studies of ancient carbonate rocks reveal three different types of cement, namely; drusy, blocky and ring cement. The drusy or drusy mosaic cement refers to elongate or acicular crystal filling or rimming voids. The crystals generally grow with their long axes perpendicular to the void margins.

Blocky cement is a mosaic of crystals in which individual crystal generally has the same diameter in all directions. These crystals tend to be granular and fill void spaces. Some crystals may appear to be the result from the growth of several drusy crystals.

Ring cement or syntaxial overgrowth are large calcite crystals, grown in optical continuity with original single crystal grain.

All the three types of cement grow, meet other crystals and also form crystal boundaries, which are not necessarily in conformity with any particular crystallographic plane and therefore are termed as compromise boundaries. Different types of cements have been presented in Fig. 10.2.

Ancient limestones indicate that cementation has taken place in several different diagenetic environments and also over a long period of time. The principal diagenetic environment for carbonate sediments is interpreted as intertidal, shallow subtidal, deeper water, meteoric and burial environments.

Intertidal cementation: As stated earlier, beach rock is produced due to cementation of beach sand in the intertidal zone, where aragonite and high Mg calcite are more common. The formation of beach rock results due to either of the following processes:

I. Physicochemical precipitation through evaporation of sea water and loss of CO_2 mainly at the time of low tide,

II. Biochemical precipitation due to decomposition of organic matter, mainly algae, and

III. The precipitation from fresh groundwater or sea water mixing with fresh water.

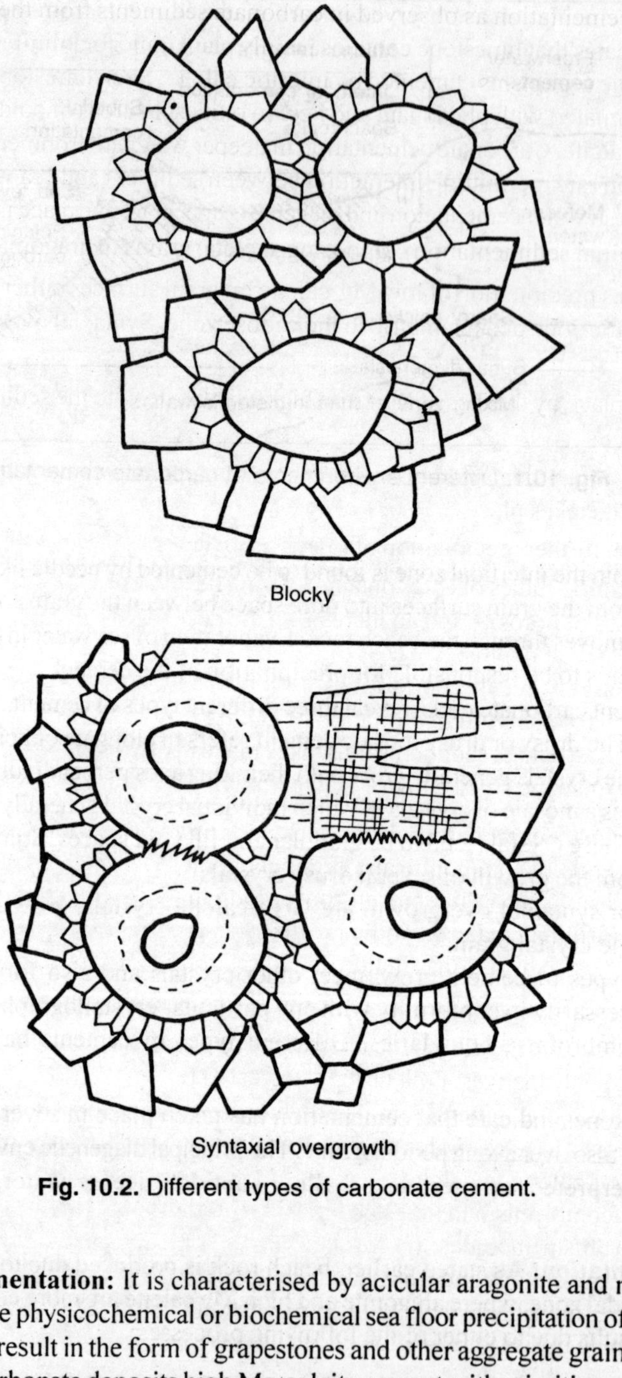

Fig. 10.2. Different types of carbonate cement.

Shallow subtidal cementation: It is characterised by acicular aragonite and micritic high Mg-calcite cement. It can be physicochemical or biochemical sea floor precipitation of cryptocrystalline aragonite, which may result in the form of grapestones and other aggregate grains.

In modern reefs, carbonate deposits high Mg-calcite cement with micritic peloidal and bladed fibrous crystals dominating over acicular aragonite. The cement is precipitated within the primary cavities between and within the framework of other skeleton. Crystals preferentially grow upon a substrate of the same mineralogy, having syntaxial relationships.

Deeper water cementation as observed in carbonate sediments from the ocean floor at depths upto 3500 m. indicates that limestone contains mainly planktonic foraminifera, mollusca, cocoliths and other benthonic organisms cemented by micritic calcite. Such limestones are generally bored and may be impregnated with phosphate and Fe-Mn oxides. In areas of cold deep water, cement is micritic low Mg-calcite. Carbonate cementation in deeper water environment is mainly due to very slow sedimentation rate, permitting interaction between sediment and sea water.

Meteoric i.e. fresh water cementation and diagenesis appear to take place in five stages as follows:

Stage I- It is the initial sedimentation composing skeletal grains of aragonite and high Mg-calcite.

Stage II- It involves precipitation of low Mg-calcite on grain surfaces either in the form of uniform fringe in the phreatic zone or asymmetric in the vadose zone. Syntaxial overgrowth may take place in this stage.

Stage III- It takes place by the loss of Mg from high Mg calcite with the sediment of low Mg-calcite and aragonite.

Stage IV- It is the main diagenetic process, wherein aragonite dissolution and reprecipitation of drusy, sparry calcite takes place.

In the last stage, further precipitation of low Mg-calcite takes place to fill the remaining voids.

Burial Cementation: Studies of cores obtained during deep sea drilling indicate that with increasing depth,there is a loss of porosity and an increase in the degree of lithification from an ooze to chalk and then to limestone at depths in excess of 500-600 m.

Cementation largely results due to dissolution of less stable fine-sized calcite crystals of pelagic skeletal material and reprecipitation of calcite upon larger crystals of skeletal origin.

Cementation of carbonate sediments thus takes place in different environments with the change in mineralogy and also the change in original texture of carbonate sediments.

Replacement: Primary carbonate rock particles, such as; ooliths, structures,such as; stromatolites and unstable minerals like aragonite and original characters, which are preserved,are termed as ghosts. Such inclusions as pseudomorphs or reproduction of an original outline to preserve the evidence of the original material indicate that the growth of new mineral has taken place simultaneously with step by step dissolution of the original mineral and precipitating another material at the same time.

Neomorphism: Some of the diagenetic processes bring about changes in mineralogy and fabric of the sediments.To such changes, Folk (1965) suggested the term "neomorphism", which suggests the transformation of one mineral into its polymorph. Neomorphism takes place by either of the following processes: (*i*) wet polymorphic transformation of aragonite to calcite, or (*ii*) wet recrystallisation of calcite to calcite.

Processes of neomorphism in limestone are of aggrading type, as a mosaic of coarser crystals is formed. Neomorphism includes: (*i*) formation of microspar in calcilutite; (*ii*) calcitization of original aragonitic skeletons and (*iii*) replacement of acicular cement by fibrous calcite.

In degrading neomorphism, development of a mosaic of finer crystals takes place.

Neomorphism gives rise to formation of calcite spars, which have the following characteristics:

(*i*) irregular or curved intercrystalline boundaries, usually with embayments;
(*ii*) irregular crystal size distribution and patchy development of coarse mosaic,
(*iii*) gradational boundaries to areas of neomorphic spar and
(*iv*) inclusions of skeletal grains in coarse spar.

In aggrading neomorphism, growth of new crystals takes place at the expense of others, generally by syntaxial precipitation on earlier crystals.

Neomorphic replacement of cement and skeletal materials takes place where a thin fluid film less than a few microns moves through the grainsor cement. Such a fluid film generally dissolves the carbonate aragonite material in front, precipitating calcite behind. In such replacement, original features of the carbonate are retained, but are unrelated to the fabric of that of calcite.

Stylolites: These are thin seams of clay and insoluble material and run mostly parallel to bedding. They are commonly sutured, while others are less regular. Stylolites form by pressure solution, dissolution of the limestone along planes, as a result of overburden or tectonic pressure. The clay that fills the stylolite seams is the insoluble residue derived from limestone solution. Stylolites are less well-defined solution planes, sometimes referred to as flasers and commonly develop preferentially in clay-rich horizons. Solution is generally common at limestone-shale junctions and also along bedding planes. Pressure solution therefore is cited as one of the principal source of $CaCO_3$ for limestone cementation in late diagenesis.

10.6 DOLOMITIZATION AND DEDOLOMITIZATION

The term dolomite refers to the mineral species, with chemical composition $CaMg(CO_3)_2$ and also to the carbonate rock that contains more than 50% of the mineral. Dolostone is a rock that is composed of the mineral dolomite.

Dolomite has an ordered, three-layered arrangment of Ca, Mg and CO_3 ions, analysis of oxides, being 21.90% MgO, 30.4% CaO and 47.70% CO_2 by weight. The most common and abundant substitution is that of ferrous ion for magnesium. Subsequent Mg^{++} and Fe^{++} dolomite is called as Ankerite, having the composition, as, $Ca(Mg, Fe)(CO_3)_2$. However, no naturally occurring CaFe $(CO_3)_2$ is known. Dolomite is found to occur in rocks of all geological ages, i.e. from the Late Precambrian to Holocene. Chemical analyses of about 9000 dolomite samples from Russian Platform and N. America indicate that Calcite/Dolomite ratio is rather infinity in the Quaternary period, that decreases through geologic time, as 80 : 1 in Cretaceous rocks to 3 : 1 in Palaeozoic rocks to 1 : 3 in Late Precambrian rocks.

These results indicate that there were increased opportunities with time for carbonate rocks to become dolomitized and also reflecting a greater prevalence of conducive depositional and diagenetic environments for the formation of dolomite in Late Precambrian and Palaeozoic times.

Requirements for Dolomite Formation

Spontaneous Nucleation: There is an apparent lack of dolomite whitings in natural solutions, indicating whether such waters are saturated with respect to dolomite. No such crystals are known to occur, thereby suggesting undersaturation. However, detrital dolomite in the form of perfect rhombohedra in the Gulf of Mexico and off the coast of South Florida has formed irrespective of long periods of time in sea water. Detrital nature of dolomite has been inferred by ^{14}C dating. Apparently, it appears that dolomite neither precipitates nor dissolves in sea water, suggesting that sea water is approximately at saturation so that dissolution or precipitation reactions are imperceptibly slow. However, sythesization of dolomite from sea water has not yet been accomplished.

Dolomite having highly ordered structure, formation is a slow process at low temperature. In the presence of Ca and CO_3 ions in solution saturated with respect to calcite, dolomite formation is generally prevented to take place.

Dolomitization of Calcium Carbonate

At low temperatures, diffusion of ions in crystals is extremely slow process, even when considered by geologic time scale. Hence, dolomitization is not regarded as solid state reaction. It occurs therefore by simultaneous dissolution of $CaCO_3$ and precipitation of dolomite from an aqueous solution, passing through the rock. In general, therefore, there is a need to meet two requirements for dolomitization to take place.

(*i*) the a.Mg^{++}/a.Ca^{++} ratio of the water must be sufficiently high to allow the reaction to take place; and

(*ii*) a mechanism, capable of flushing in sufficient volumes of dolomitizing fluids through the calcium carbonate rock for the reaction to be completed to form dolomite rock.

Mechanism of Dolomitizatin

Evaporative Reflux: In the Permian Reef complexes of West Texas and New Mexico, association of dolomite in front and under the evaporite sequence is suggestive of the concept of evaporative reflux.

It is suggested that the evaporation of sea water to the point of precipitation of gypsum, i.e. $CaSO_4 \cdot 2H_2O$ at a concentration slightly in excess of three times sea water would remove calcium from water; thereby automatically raising Mg/Ca activity ratio above 5.6 found in sea water. This process of evaporation gives rise to denser water that tends to sink down through earlier deposited sediments and rock, pore spaces of which were earlier filled with lighter sea water. Flow velocity of this water depends on the density contrast and permeability of the underlying material. Thus, a water having a high a.Mg^{++}/a.Ca^{++} could be flushed in large volumes through earlier deposited limestone (Fig. 10.3). Adams and Rhodes (1960) have termed this process as 'seepage refluction'.

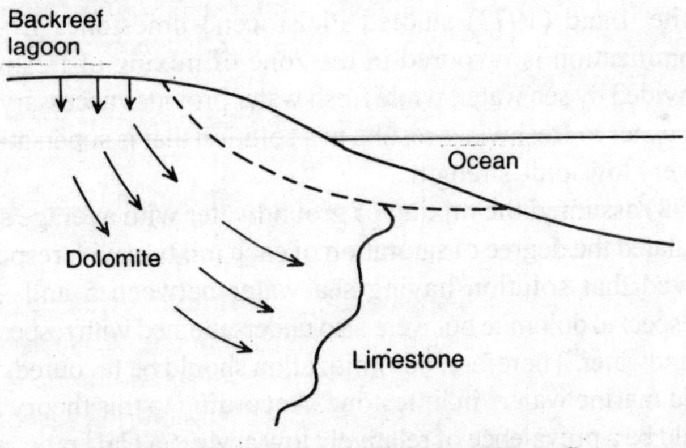

Fig. 10.3. Mechanism of evaporative reflux.

An example of modern dolomitization has been reported from the southern end of the Bonnaire island in the Netherlands Antilles. It is found to be associated with gypsum concentration. The main lake - Peckelmeer - deposited normal marine sediments before the full development of Barrier Ridge had taken place. In the last hundreds of years, lake has accumulated gypsum. Sea water enters the lake through springs in most of the year, as the lake is below sea level. In spite of the fact that sea water is constantly added to the lake, it never reaches to the stage of concentration to give rise to the formation of halite. This indicates that evaporation is constantly taking place from the surface of the lake and also from the pore surface of the surrounding flats. Such a condition can be explained only by a method of removing brine from the lake, for which the only possible exit route appears to be through the underlying sediments and rocks. The South Bonaire island, therefore, provides an example of modern dolomite in association with a brine, that gets concentrated by evaporation beyond the point of precipitation of gypsum; and precipitation of gypsum causes the increase in Mg^{++}/Ca^{++} molar ratio from 5.2 in sea water to as high as 20 or more in the brine. Such a concentrated brine is flushed through the underlying sediments and rocks. This mechanism that provides the condition favourable for dolomitization is termed as 'Evaporative Reflux'. Radiocarbon dating of this dolomite deposit indicates that it is modern and has been formed probably within the last 1000 years.

Though the evaporative relfux is a process responsible for dolomitization, dolomite deposits are not found to be associated with evaporites, i.e. evaporite beds; such as gypsum deposits. A reason might be that evaporites are easily dissolved and therefore much of the dolomite in the geologic record is not found to be associated with evaporites; i.e. gypsum deposits.

Dorag Dolomitization

Dolomite as a mineral is a precisely ordered form and consists of layers of carbonate ions that alternate with layers of Ca^{++} and Mg^{++} ions due to which they are not easily segregated into separate layers. Folk and Land (1975) are of the opinion that segregation of ions to produce dolomite can be accomplished rather at a faster rate, if the rate of crystallisation is very slow or if the solution is dilute with few interfering ions. According to them, it is easier to form minerals at the slow rate of crystallisation, because then the minerals are in thermodynamic equilibrium with the surrounding solution and precise Mg : Ca ordering could be obtainable from solutions with Mg/Ca activity ratio near the theoretical value of 1 : 1.

Fresh or nearly freshwater provides an environment of low salinity and slow rates of crystallisation. Both the surface and subsurface freshwaters usually have too low Mg/Ca activity ratio to permit the formation of dolomite. Land (1973) studied Pleistocene limestones from Jamaica with the conclusion that dolomitization is favoured in the zone of mixing of freshwater and sea water, magnesium being provided by sea water, while fresh water provides necessary dilution. Addition of small amounts of sea water to freshwater results in a solution that is supersaturated with respect to dolomite and with a very low ionic strength.

Badiozamani (1973) assumed the mixing of groundwater with average sea water in different proportions. He calculated the degree of saturation of each mixture with respect to both calcite and dolomite and observed that solution having sea water between 5 and .30% were not only supersaturated with respect to dolomite but were also undersaturated with respect to calcite, compared with the original groundwater. Therefore, dolomitization should be favoured, where there is a zone of mixing of fresh and marine waters in limestone. According to this theory for dolomitization to take place, there should be a prevalence of relatively low $a.Mg^{++}/a.Ca^{++}$ ratio, a mechanism to move the solution through limestone and a sufficient time for the process to take place. For the purpose, a long-term stability of groundwater system provides sufficient time for dolomite growth, even if Mg^{++}/Ca^{++} activity ratio is greater than unity.

For such a process, the term Dorag Dolomitization has been suggested by Badiozamani (1973) (Fig.10.4). Such a process is independent of gypsum precipitation and high Mg^{++}/Ca^{++} ratio and therefore can be accounted for dolomites that lack evidence of association with evaporites.

The process of dolomitization, therefore, takes place during early or late stage of diagenesis.

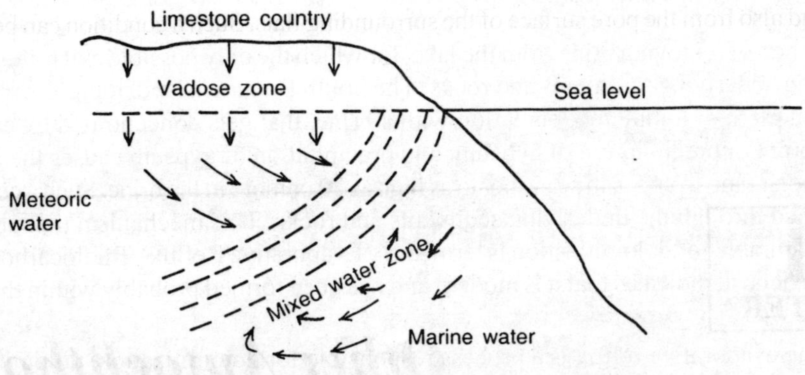

Fig. 10.4. Mechanism of dorag dolomitisation.

10.7 DIAGENETIC EFFECTS

The preceding paragraphs have dealt with the various processes that are post-depositional and bring about various diagenetic effects in limestone. Main diagenetic processes in limestone include drusy crystallisation, neomorphism, dissolution or leaching and silicification. These processes bring out, apart from the changes in framework structure, changes in porosity of limestone.

In drusy crystallisation, development or growth of calcite or aragonite crystals takes place in voids or open spaces, thus resulting with the decrease in primary porosity.

Neomorphism takes place by development of new calcite crystals or calcite spars at the expense of cement. This process therefore takes place without either creating or destroying voids or open spaces. Hence, there is no change in the primary porosity. However, if neomorphism takes place by the process of polymorphism, there is a definite porosity response to polymorphic changes. Thus, if the earlier formed aragonite inverts to calcite, there is a decrease in porosity by about 8%. If calcite is replaced by dolomite, i.e. the process of dolomitization, there is nearly 13% increase in porosity; however, in the process of dedolomitization; i.e. dolomite being replaced by calcite, there is decrease in porosity by about 13%

The process of leaching or dissolution is a common process in limestone country, that takes place always with an increase in porosity, which is more commonly of moldic and vuggy types.

Silicification is also a common post-depositional, diagenetic process in limestone. It takes place either by replacement or by chalcedonic pore filling. If the silicification takes place by replacement, no change in porosity takes place. However, if it is due to pore filling, there is a reduction in primary porosity.

DEDOLOMITIZATION

This process can take place, when dolomite comes in contact with meteoric waters; wherein replacement of dolomite by calcite results to produce limestone again. Calcite replacement of dolomite can also take place in association with gypsum anlydrite solution, which is a near-surface phenomenon. Dedolomitization can be recognised by noting shapes of dolomite crystals, usually rhombohedra, occupied by calcite as pseudomorphs or calcite crystals with replacement fabric, containing small inclusions of dolomite. In some cases, layers and concretions of fibrous calcite randomly yet completely replace the dolomite or even original limestone texture is partially regenerated on dedolomitization.

11
CHAPTER

Other Autochthonous Rocks: Evaporites

11.1 INTRODUCTION

The term evaporites indicates sedimentary deposits that consist of minerals which are precipitated from brines by evaporation. These are a group of authigenic rocks containing minerals formed by the process of evaporation. Gypsum ($CaSO_4 2H_2O$), anhydrite ($CaSO_4$) and halite (NaCl) are by far the most common evaporite minerals. Evaporites have been reported practically from all the continents. According to Pettijohn (1962), nearly 25% of the continental area is covered by these rocks. Majority of the evaporite deposits are of Phanerozoic age. Their formation during the Precambrian period has also been reported from some of the continental areas.

11.2 GENERAL CHARACTERISTICS

Formation of evaporites is controlled by a certain conditions of geochemical parameters of water within the basin. These parameters are temperature, salinity, acidity-alkalinity and proportions and kinds of dissolved gases, such as; oxygen, carbon dioxide, hydrogen sulphide and methane, which decipher density of water. Density stratification results and which, in turn, brings out either circulation or there is a lack of circulation. In case, where salinity in water reaches to high proportions, precipitation of evaporite minerals takes place. Such a precipitation generally results in brines and is the product of natural inorganic chemical reaction. It has been observed that a definite interrelationship exists between progressive increase in concentration of brines and precipitation of evaporite minerals.

The brine from which these minerals precipitate, form by one of the following ways, from sea water or by reconstructed brine. For reconstructed brine, rain water or groundwater is allowed to pass through rock that contains soluble salts and calcium-sulphate minerals. This meteoric water flows down, and becomes a part of the groundwater system. It then comes out in nearby valley floors. Evaporation of this brine in the valley floor sediments or in playa lakes increases the concentration of brine to the point at which gypsum can precipitate.

It is observed that evaporation of a 1000 m. column of normal sea water can produce only 0.75m. thick of gypsum and about 13.5 m. of halite. Precipitation of gypsum starts when sea water has reached concentration to slightly more than one third of its original volume, whereas halite begins, to precipitate, when sea water has reached nearly 1/10th of its original volume by evaporation. Association of thick deposits of gypsum or anhydrite with little halite implies that there was a continuous source of new sea water getting added to brine for evaporation and that the brine rarely reached the point of halite precipitation.

The formation conditions of gypsum and halite can be explained by the following experiment. When a container of sea water is allowed to evaporate and new sea water is constantly added to maintain original volume, precipitation of halite takes place. But, if the concentrated brine is removed at the same time, as new sea water is added, then the concentration is maintained at the equilibrium and deposition of gypsum results (Fig.11.1).

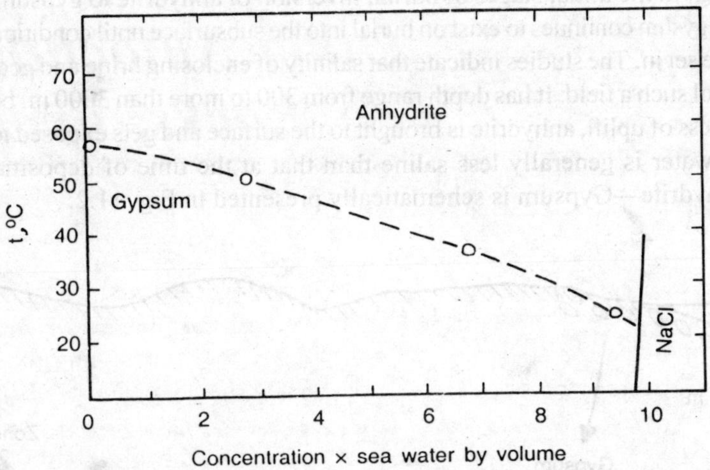

Fig. 11.1. Gypsum–anhydrite equilibrium diagram.

Another mechanism is the subsurface reflux that produces same results. In this mechanism, supply of new sea water is made either by surface flow or seaway or through sea water springs. The concentrated brine is removed down through the sediments and rocks, which underlie and also surround the evaporite basin. The flow of concentrated brine water results due to density difference between it and the lighter water, which is present in the porous spaces within the sediments and rocks. The similar mechanism is also applicable to concentration of brine within the vadose zone or phreatic zone.

Arid and hot climate is considered to be responsible for the formation of evaporites. For the purpose, rate of evaporation should be sufficient to cause significant concentration of brine and to counteract the inflow of new water and loss of brine water from the system. Generally, the conditions of high rate of evaporation are encountered in areas of high temperature and low rainfall. However, deposition of gypsum has also been reported in the present day from Arctic and Antarctic regions.

11.3 GYPSUM-ANHYDRITE CYCLE

Calcium sulphate minerals occur as individual crystals or cluster of crystals in the host carbonate or detrital rocks. This mineral forms a major part of layered rock with minor amounts of carbonates, siliceous or carbonaceous material. In nature, there are generally two minerals, namely; gypsum ($CaSO_4{\cdot}2H_2O$) and anhydrite ($CaSO_4$). The minerals containing less amount of water molecules, such as; bassenite or calcium sulphate hemihydrite are rather unstable and convert immediately to gypsum in the presence of water.

In nature, a cycle of calcium sulphate minerals is found to exist. Such a cycle results stability relationship between calcium sulphate-water system and also the normal sequence of deposition —burial—uplift—erosion. The first mineral to form in the cycle seems to be gypsum. It is deposited in an evaporite basin, where water is evaporated to the point of gypsum precipitation. Similar process of formation of gypsum operates even within the vadose or phreatic zone below the tidal flats. Formation of gypsum within the vadose or phreatic zone is mainly due to the replacement of the earlier deposited detrital or carbonate sediments. The gypsum so formed may get replaced by anhydrite in near surface conditions.Anhydrite can persist on burial until the anhydrite stability field continues to exist. The stability field is defined by temperature and pressure and activity of associated water. In the initial stages of burial, inversion of anhydrite to gypsum is possible. The initially formed gysum continues to exist on burial into the subsurface until conditions of replacement by anhydrite are set in. The studies indicate that salinity of enclosing brine and geothermal gradient generally control such a field. It has depth range from 300 to more than 3000 m. below the surface. Due to the process of uplift, anhydrite is brought to the surface and gets exposed to erosion. At this stage, ground water is generally less saline than that at the time of deposition. The cycle of Gypsum—Anhydrite—Gypsum is schematically presented in Fig. 11.2.

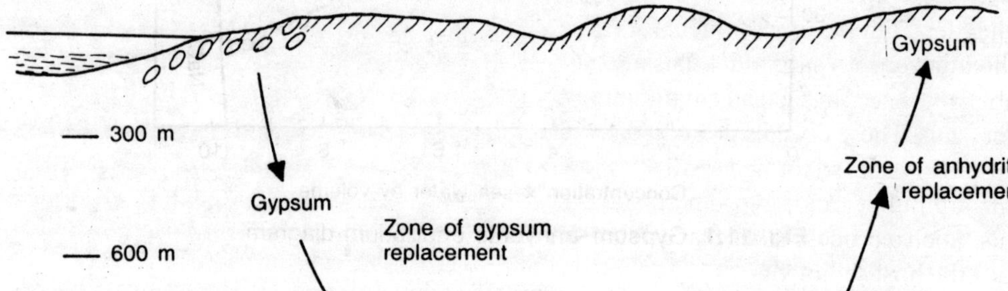

Fig. 11.2. Gypsum–anhydrite–gypsum cycle.

11.4 DEPOSITIONAL ENVIRONMENTS

As stated earlier, evaporite minerals form in evaporite basin and also in phreatic or vadose zone in the subsurface by the process of evaporation, when concentration of brine reaches to the point of precipitation of these evaporite minerals.

Isolated water bodies: Deposition of evaporites can result in nearly enclosed marine basins. Formation of such a basin may take place by two important methods; such as:

(*i*) the process of rifting, accompanied with opening of ocean basin and formation of the carbonate platform with associated starved basins, which later have been filled with evaporites. Such conditions appear to have resulted in the past. At present, however, there are no such huge isolated basins or water bodies in which deposition of evaporites is taking place. However, to create such conditions for deposition of evaporites may be relatively uncommon; the reason for which appears to be the degree of restriction needed for the existence of isolated water bodies. For an extensive evaporite deposit to form, there is a need of passage for new sea water to enter continuously into the isolated water body. Such an entry should have small surface connection to the isolated water body that is totally cut off at the surface and is fed only by sea water spring. Fig. 11.2 is prepared after considering the present surface channel to the sea, their surface area and salinities.

(*ii*) Another important factor for evaporites to deposit is the permeability of underlying sediments.If the permeability is high, the subsurface reflux of brine takes place at a relatively low concentration and water body therefore never reaches a point of concentration at which gypsum can deposit. In some ancient evaporite basins, complete surface isolation of the basin appears to have taken palce and sea water was supplied by seepage through the sediments that emerged as a spring within the evaporite basin. Such a process of supply of sea water through springs seems to be advantageous over surface channel supply, as it provides favourable conditions constant for longer period of time. This process explains the formation of an uninterrupted thick accumulation of evaporite deposits.

Another depositional environment for calcium sulphate mineral is within the vadose zone and upper phreatic zone on supratidal flats and also in desert playas. For such sediments, a common term 'sabkha' is used. In such an environmental set-up, gypsum grows within the sediments in the form of crystals. These crystals grow by displacing sediments, provided sediments are sufficiently soft. Gypsum crystals so formed are free of inclusion of any sedimentary particle. However, if the sediments are rather rigid, gypsum crystals grow in pore spaces commonly with inclusions of sediments. Such a process takes place as the force of gypsum crystallisation is not strong enough to displace the host sediments

In the vadose zone, air and water are in contact with each other. Evaporation in this zone results in the brine concentration, which gradually becomes supersaturated with respect to gypsum. Such a concentrated brine sinks down or reflux down through the underlying sediments and sometimes even below the water table. Until the brine remains supersaturated, gypsum continues to deposit.

When the regressive phase is set in, the intertidal zone shifts seaward. In such a setting, a vertical section of earlier marine to supratidal sediments gets exposed as the supratidal flat. Gypsum crystals that grow from the concentrated brine are now seen within these sediments. These crystals had grown in the earlier intertidal and marine sediments

Such gypsum crystals, formed by displacement of sediments are buried further. If P-T conditions for anhydrite formation prevail, gypsum is replaced by anhydrite. Replacement of gypsum by anhydrite takes place with nearly 38 percent loss of volume, which is in the form of nodular crystals of anhydrite.

11.5 NATIVE SULPHUR

In common association with evaporite minerals, such as gypsum-anhydrite, sedimentary sulphur deposit can also be encountered. An important process in the sedimentary environment is the chemical reaction, wherein reduction of sulphate mineral gives rise to hydrogen sulphide and free sulphur.

Formation of sedimentary sulphur can take place, where evaporites contain gypsum or anhydrite. Such sedimentary sulphur deposit commonly occurs in limestone units. Extensive deposits of sedimentary sulphur have been reported from Permian carbonate rocks in USA. These deposits occur predominantly in zones of porosity and are associated with earlier oil traps. Hydrogen sulphide released from petroleum might be a partial source for sulphide formation. Sulphur deposit rarely occurs in the form of continuous bed.

Field observations reveal that sedimentary sulphur has formed from calcium sulphate, i.e. gypsum or anhydrite. Porous limestone that overlies gypsum or anhydrite serves as a good host for thick accumulation of sulphur.

Formation of sulphur has been generalised with the help of following equation,

$$6CaSO_4 + 4H_2O + 6CO_2 \rightarrow 6CaCO_3 + 4H_2S + 2S + 11O_2$$

This equation indicates the reduction of sulphate to sulphide. Such a process of reduction brings out large isotopic fractionation with the sulphide, gradually getting enriched in lighter isotopes. This fractionation is similar to the one that has been experimented in the laboratory in the presence of sulphate reducing bacteria. This bacteria most commonly is Desulphovilbrio desulphuricans. It has been observed that these bacteria can produce H_2S at a maximum rate of 1000 ppm per day. It has, therefore, been suggested that the sulphur alongwith limestone cap rock are produced by sulphate reducing bacteria that used petroleum and brings out conversion of gypsum or anhyrite to sulphur and calcite.

12

CHAPTER

Chert

12.1 INTRODUCTION

Chert is classified as an authigenic intrabasinal rock. It is a fine grained siliceous sediment of either chemical or biochemical or biogenic origin. It is tough and brittle rock composed of pure silica. It exhibits splittery to conchoidal fracture and vitreous lustre. Common varieties of chert are jasper, flint, novaculite and porcellanite.

On the basis of occurrence of chert in various stratigraphic horizons, it is classified broadly into two groups. as; nodular chert and bedded chert. Bedded cherts are generally associated with volcanic rocks, while nodular cherts, that commonly occur in limestone, are considered to be diagenetic in origin, and are regarded to have formed by replacement; whereas, bedded cherts are considered to be primary accumulation.

12.2 TEXTURE

Microscopic studies of chert show the following varieties, as; cryptocrystalline, medium crystalline and coarse crystalline or fibrous. The fibrous variety of chert is termed as chalcedony, which is a polymorph of quartz. It displays a pattern of radial fibres. The other crystalline varieties are quartz, porcellanite, cristoballite or opal. Cristoballite is a metastable polymorph of quartz, while opal is a hydrous silica and contains about 8 to 9 % of water.

Texture of chert can best be studied under electron microscope. These studies reveal the presence of polyhedral crystals of quartz, which are subequant to elongate in shape. Euhedral crystals of quartz are generally absent. Chert studies indicate that the minimum possible crystal size is related to the thermodynamic, i.e. physico-chemical, conditions prevalent in silica solution, from which chert has formed. These studies have indicated the inverse relation between the two.

In fine grained, microcrystalline aggregates of chert, circular or elliptical clear areas, if present, may be of radiolarians. Similarly, siliceous sponge spicules are quite frequent in spiculite chert. The fibrous aggregates show wavy extinction. Some varieties of chert are relatively impure and contain very minute carbonate particles, disseminated in the body of chert. Generally, these inclusions are predominant in porcellanite and are found to be of rhomb-shaped calcite or dolomite. The inclusions may be uniformly distributed or may occur as concentrates in laminations. Chert associated with iron bearing formations has inclusions more commonly of siderite.

Some of the rocks contain a mixture of clay or silty clay, alongwith variable proportions of opaline silica. It is more or less porcelaneous and is termed as porcellanite. These porcellanites gradually grade into siliceous shale.

Uncommon variety of chert is sandy chert. Gradations between chert, sandy chert and sandstone with cherty cement have also been reported.

Another class of chert is oolitic chert. The siliceous chert is of replacement origin. Oolites are about 1 mm in diameter and are composed of chalcedony, having concentric structure. Nuclei of oolites consist more commonly of rounded grains of quartz around which development of secondary quartz in optical continuity takes place to give rise to oolites. A boundary between detrital quartz nucleus and secondary siliceous overgrowth is marked by inclusions.

Nodular chert generally is devoid of any internal structure. It has a dense black interior, while the exterior is lighter and white in colour. Commonly, chert is fossiliferous, fossils being either calcified or silicified.

12.3 CHEMISTRY OF CHERT

The most abundant element in the earth's crust is silicon, which when combined with oxygen gives rise to silicon dioxide termed as silica. Majority of terristrial rocks are silica rich. On weathering and erosion, these rocks give rise to siliceous sediment or silica; which is then supplied through rivers into the oceanic environment. This silica is initially deposited in near-shore environment, while some of it reaches deep oceanic areas. Some of the dissolved silica gets trapped in pore water during sedimentation; whereas silica removed from seawater is used by organisms to construct their skeletons. The skelaton is opaline, hydrated amorphous,biogenic silica, and it is termed as biogenic fixation of silica. These organisms are diatoms, radiolarians and siliceous sponges. After the death of organisms, their skeletons settle through water column and most of the silica is redissolved. Dissolution, therefore, is another step in the cycle of marine silica. This is referred to as monoxidative dissolution. When such an accumulated siliceous material is greater than 30% of the sediments, it is termed as siliceous oozes. If diatoms dominate in greater than 30% proportion, they are termed as diatoms oozes and if radiolarians dominate by more than 30%, it is termed as radiolarian oozes.

Seawater is undersaturated in relation to silica, as solubility of amorphous silica is 120 ppm. Therefore, no direct precipitation of silica from seawater is possible. However, formation and precipitation of biogenic silica through skeletons of organisms seem to be a plausible mechanism. In modern oceanic environment, diatom oozes are nearly 20 % in siliceous sediments of the oceanic environment. Sponge spicules are rather subordinate in amount. It is observed that siliceous oozes occur in areas with high biological productivity and areas of upwelling and nutrient supply. Diatom oozes occur in high latitude areas of northern Atlantic Ocean, while radiolarians oozes are confined more commonly to the equatorial regions of Pacific and Indian oceans.

Deposition of siliceous oozes in oceanic environment is controlled by carbonate compensation depth (CCD). CCD is the depth above which carbonate rich sediments accumulate and below which carbonate free sediments accumulate. Mean depth of CCD is about 4500 m. (4.5 km.), which is about mid distance between crests of mid-ocean-ridges and deepest parts of the ocean basin. It has been observed that variation in CCD takes place in Pacific and also in Atlantic Ocean and which, in turn, is controlled by upwelling and organic productivity. In areas of upwelling and high organic productivity, the CCD is depressed; whereas in areas close to continents, where there is high organic productivity, CCD becomes shallower. Thus, variation in CCD also affects areas of formation and precipitation of biogenic silica.

Conversion of silica or biogenic silica to chert is a much debated problem. The problem of the formation of chert is not yet well understood, though there are many theories that account for origin of chert. Different theories have discussed the problem of silica diagenesis and formation of chert. Formation and welding of the tests of micro-organisms, is the first stage of silica diagenesis. Four stages in formation of chert in carbonate sediments have been suggested:

(*i*) infilling of empty foraminiferal chambers with chalcedony or chalcedonic quartz,
(*ii*) replacement of groundmass of the carbonate rock with fine grained cristobalite,
(*iii*) replacement of foraminiferal tests by chalcedony and chalcedonic quartz, and
(*iv*) final infilling with silica of all the voids in the rock and inversion of cristobalite matrix to quartz.

Two theories have been advocated; for the formation of chert viz., maturation theory and mineral association theory. According to maturation theory, silica diagenesis proceeds in a given time through various stages, stated above, whereas according to mineral association theory, formation of chert is related to specific sedimentological association.

12.4 ORIGIN

For the origin of chert, three sources for silica have been identified, which are mainly, volcanogenic origin of silica, diagenetic alteration of smectite to illite releasing silica and diagenesis of biogenic silica. Of these, the last one is commonly accepted theory for the source of silica and thereby the formation of chert.

As stated earlier,biogenic silica is in the form of amorphous hydrated opal. It is referred to as opal-CT, also termed as disordered cristobalite, alpha-cristobalite and lussatite. Opal-CT consists of interlayering of cristobalite and tridymite, which replaces diatom and radiolarian skeletons. Later, precipitation in the form of bladed crystals in cavities takes place. In the next stage of diagenesis, opal-CT is converted into quartz chert, as equant mosaic of macroquartz crystals and also chalcedonic quartz.

Formation of chert from biogenic opal-A results due to solubility differences and chemical conditions. Solubility of biogenic silica is in the range of 120 - 140 ppm; that of cristobalite is in the range of 25 - 30 ppm and of quartz is 6 - 10 ppm. The solubility is considered in the pH range of marine environment. As the metastable opal-A dissolves, the solution gets saturated with respect to opal-CT and quartz, and precipitation of opal-CT in preference to quartz takes place. It is because quartz being more ordered structure requires slow rate of precipitation. Role of temperature during such transformation, is also significant. With increase in burial depth, there is a relative increase in temperature that facilitates transformation of opal-CT to quartz. Transformation of opal-CT is referred to as maturation process.

It has been shown that maturation of opal-A to chert also takes place. It is however, controlled by nature of host sediment and chemical condition.

Bedded chert: It is the most common occurrence of precipitated silica. In some cases, it makes up even the entire formation. Commonly, it is found to be thick and areally extensive. On the basis of its occurence, it is classified into three main groups, as; cratonic chert, deep sea chert and those associated with evaporites.

Cratonic chert is found to be associated with limestones, generally in the form of nodules. Chert in carbonate rocks forms an anastomising, three dimensional network. With an increase in the proportion of chert, limestone gives rise to chert rocks with an intercalated bedding of limestone. Such a bedding, however, is wavy and irregular. In such limestones, silicified calcareous fossils are found to be abundant. Brachiopods and bryozones and the like are the common occurrence.

Origin of the bedded cherts in cratonic environmrnt is a very controvertial subject and many theories to explain the origin of these deposits have been suggested. These can be classified into two groups. According to one, the silica is considered to be contemporaneous with sedimentation, whereas according to the other, the silica is post depositional replacement of the host rock. Although, there appears to be no general agrement regarding the origin of chert, most researches of bedded cherts consider them as precipitates of silica gel (Carozzi, 1960; Pettijohn 1969). According to Dapples (1967), silica of the bedded cherts was deposited contemporneously with the clastic sediments or that it was precipitated immediately after the deposition of detritus. However, to account for the large amount of silica needed to form chert layers, Dapples (op.cit.) has opined that, "there appears to be in current favour the concept that the localised volcanism may have supersaturated waters sufficiently to result in precipitation of the required enormous amount of silica. According to such a concept, the silica is precipitated as a primary constituent of the sediment and may under circumstances constitute the entire sediment being accumulated".

Deep sea chert is layered chert and rhythmically bedded. Layers are a few cm.in thickness and are generally separated by thin partings of siliceous or sideritic material. These bedded cherts, on closer examination, are found to bifurcate, and in these layers, it is generally cryptocrystalline, dense and sub-vitreous in lustre.

Chert associated with evaporites are nodular and bedded, nodules being marked by a reticulate pattern. It generally occurs as interbedded siliceous claystones and is found to be associated in some cases with dolomites.

Association of bedded chert with some volcanic rocks also has been reported. In such cases, chert is found to be deposited either below or above the lava flows. Lava flows and volcaniclastic sediments are found to be intercalated.

Presence of chert in limestone of fresh water lacustrine origin has been reported by Joshi (1993). These are the cherty Lameta Limestones of Maestrichtian age. On the basis of the field and petrographic studies, he has inferred the origin of chert as secondary. Its occurrence in the upper layers of limestone, has been ascribed to be either due to replacement and/or infillings and silica bearing solutions from Deccan basalts are the source, whereas its presence in the lower layers is because of dissolution of detrital quartz grains due to change in pH conditions during diagenesis. Occurrences of bedded chert have also been encountered along the tectonically active plate margins. In association with chert are graded turbidites, ophiolites and also melange. However, each type of bedded chert is not necessarily associated with ophiolites or melanges, and therefore depositional setting for such bedded chert is not certain.

Presence of thin chert beds in non marine sequences has also been reported. Such an occurrence is interpreted as lacustrine, because of the absence of marine fossils and presence of evaporite minerals. Chert in such an environment is present in the form of thin discontinous beds or as plates or as nodules. It is generally formed just below the water-sediment interface. Most of the chert is chalcedonic. Characteristic deformational features are found to be associated with such chert beds. Such features are slump structures, intraformational breccias, casts of mudcracks, formed mainly because of shrinkage during the transformation of precipitated sodium silicate to chert. In lacustrine environment, pH is gradually lowered due to influx of fresh water. Biogenic production of carbon-dioxide, associated with gradual concentration of lake water due to evaporation, silica is precipated as sodium silicate mineral-magadiite. Formation and precipitation of chert are thus preceded by the precipitation of hydrated sodium silicate mineral.

Chert is a mineralogically simple rock, however, though it is found to originate by different processes and occurs in different stratigraphic and tectonic environment.

13

CHAPTER

Phosphates

13.1 INTRODUCTION

Phosphates or phosphorites are the important natural resource. These are the chief constituents of fertilisers and have also the applications in chemical industry. Phosphate minerals occur in sedimentary rocks that range in age from Precambrian to Holocene. These phosphate deposits are also mined, apart from fertilisers, for uranium, vanadium and rare earths. These are also considered as potential source rock for oilfields.

As it serves as a primary nutrient in marine environment, organic productivity is controlled by phosphates. It occurs in two states, as dissolved 'orthophosphate' and 'particulate phosphate'. Orthophosphate is utilised during organic productivity in the upper levels of oceans through phytoplanktons. Nodular phosphorites can form in oceanic environment between a few tens of meters and even upto thousands of meters. In continental shelf, miogeoclinal and eugeoclinal tectonic settings, bedded phosphate is generally formed.

13.2 MINERALOGY

Sedimentary phosphate minerals are chiefly the varieties of apatite. These varieties occur in isomorphous series with the following end members:

$Ca_5(PO_4)_3F$ fluorapatite
$Ca_5(PO_4)_3Cl$ chlorapatite
$Ca_5(PO_4)_3OH$ hydroxyapatite

The fluorapatite is common in igneous rocks, whereas, in sedimentary environment, replacement of phosphate by carbonate may take place, while fluorine may be replaced by hydroxyl or chlorine ions. Even calcium ions may be substituted by sodium, strontium, uranium and rare earth elements. Fluorapatite generally contains upto 10% carbonate ion and is commonly referred to as francolite, having more than 1% fluorine, which is most common in sedimentary rocks. With carbonate hydroxyapatite and less than 1% fluorine, it is termed as dahllite.

13.3 ANCIENT AND MODERN DEPOSITS

The first occurrence of phosphorites or phosphate deposits on sea floor was reported by 'Challenger' oceanographic expedition in 1870s, from the continental shelf and slope off Southern

Africa. Since then, extensive nodular phosphate deposits are now known to occur on the sea floor at both shallow and deeper depths, from many areas of the world. Prominent areas are west coasts of north and south America and the continental shelves off north west Africa and Japan. These are the nodular deposits and occur in areas of very slow rate of clastic sedimentation, on the outer continental shelves and slopes, on fault scarps and flanks of submarine canyons, and generally occur between the depth range from 60 m. to 300 m. The phosphate nodules and slabs are brownish in colour. Nodules have flat base and are several centimeters in diameter, even reaching at times a meter or more across. Internally, nodules are either homogeneous or concentrically laminated or conglomeratic and contain pellets and ooids. However, many nodules exhibit layering that is irregular and in nonconcentric manner. Layers vary in thickness from few mm. to a few cm. Mineralogically, collophane is a principal mineral, while francolite and dahllite are also present. Frequently associated are glauconite and ferromanganese oxides. Some of the nodules contain more than 95% phosphate minerals. In India, one of the richest and extensive bedded deposit of phosphorite occurs at Zhamarkotra in Udaipur District, Rajasthan. It is of Precambrian-Arvalli age.

Relatively rare but modern accumulation of rich phosphatic substances is Guano. It is an excreta of animals, formed by penecontemporaneous alteration. Due to leaching of fresh guano, insoluble residue of calcium phosphate is formed. Thick accumulation of bird guano have been reported from some oceanic islands in the eastern Pacific Ocean. Geologically, however, guano is not considered significant.

Phosphate deposits or phosphorites are generally related to the areas of upwelling current and high organic productivity. The ancient deposits are of bedded nature and nearly all are considered to be of marine origin, mainly due to the presence of calcareous organisms of shelf habitat, remnants of reef building algae, cross-bedded structure and other similar characteristics of shallow marine origin. Carbonate fluorapatite is the dominent constituent mineral of phosphorite that occurs as microcrystalline cement, as oolite and pellets, bioclasts and intraclasts. Phosphorite deposits exhibit all the similar characteristics as that of limestone deposits. In other words, these are phosphoritic limestone and dolomite, bioclastic phosphorite or phosphatic mud rock. The ancient bioclastic and pelletised phosphorites are related to current upwelling, as these deposits occur on palaeo-continental margins of Tethys ocean.

13.4 PHOSPHATE GENESIS

Origin of marine phosphorites is considered on similar lines as that of dolomites. It is considered to be either primary, deposited by inorganic precipitation or it is of replacement origin. The factors that control the formation of phosphorites are current upwelling and low sedimentation rates.

It has been noted that modern phosphorite deposits are located in warm climate between the latitude 40° on either side of the equator. In these areas, upwelling currents bring cold, nutrient rich waters to the surface (while surface waters are diverted to offshore because of Coriolis effect). Upwelling currents are responsible for high organic productivity and growth of phytoplanktons in surface waters. This gives rise to organic-rich and therefore, phosphate-rich sediments and oxygen deficient waters on the sea - floor. It has been noted that P_2O_5 content of sea water is maximum at depths, between about 30 and 500 m. At shallower depths, phytoplanktons during photosynthesis rapidly consume phosphate, while at greater depths, due to increased CO_2 pressure, saturation of sea water with respect to apatite is prevented. Precipitation of solid phosphate takes place on the floor of shelves between the depths of 50 and 200 m. Process of such an inorganic precipitation of phosphorite explains the formation of modern phosphorite deposits, that are latitudinally controlled. However, this process does not explain the rarity of occurrence of phosphorite deposits during the last few million years or so. Answer for such an absence seems to lie in the equillibrium between organic and inorganic complex processes involved in formation of phosphorite.

Though, it has been argued that formation of phosphorite takes place by direct precipitation from sea water,its occurrence as some type of colloid also suggests its formation by the process of metasomatism. As stated earlier, structures and textures and field occurrence of phosphorite deposits resemble much to those of limestone deposits. It has also been found that calcareous fossils and ooliths exhibit all the stages of replacement by apatite, in phosphorite deposits. These evidences lead to suggest metasomatic i.e., replacement origin of phosphorite. However, it seems to be rather difficult that entire bedded deposit can form just by the process of replacement, while ruling out the direct precipitation from marine water or accumulation of original phosphatic materials.

Formation of extensive marine phosphorite deposits is achieved by the process of reworking. Oceanic currents remove much of the fine, unphosphotised sediments from the sea floor, resulting into the concentration of nodules, pellets and coprolites. These are under various stages of induration, which are then further phosphotised. Even greater reworking of phosphatic sediments takes place in response to changing current pattern or sea level, i.e. transgressive - regressive events, affecting the continental shelf-slope areas. Such an extensive reworking brings out upgrading of the sediments into a high-grade phosphorite deposit.

From the preceding discussion, following sequence in the formation of phosphorite deposits can be established;

(*i*) in warm and arid climatic zones and in areas of slow clastic sedimentation rates, upwelling currents are responsible in bringing nutrient rich waters;

(*ii*) formation of abundant biota due to high organic productivity;

(*iii*) death of organisms leads to anoxic organic rich sediments;

(*iv*) leaching of phosphate from the remains of organisms at the sediment-water interface, giving rise to phosphorous rich interstitial waters;

(*v*) formation of phosphorite deposit by direct-inorganic precipitation through interstitial waters or phosphotisation of sediments;

(*vi*) reworking and upgrading of sediments into high-grade phosphorite, in response to changing current pattern or sea-level, i.e. transgression-regression.

14

CHAPTER

Iron and Manganese Rock

14.1 INTRODUCTION

Iron is one of the most abundant elements in the earth's crust. It is present in substantial quantities, either in the form of oxide or silicate, in terrigenous clastic rocks. Sedimentary rocks that contain more than 15 % Fe are considered as iron rich.

The iron-rich sedimentary rocks host the largest reserves of iron-ore of the world. These are mostly Precambrian Banded Iron Formation, referred to as taconite.It contains on an average 25 to 40 % Fe. Extensive and varied types of iron- deposits have been reported from geologic record but their modern analogs however, are lacking. Two groups of iron-rich rocks belonging to two different geologic periods have been identified, as; the Precambrian Iron Formations and the Phanerozoic ironstones. Major difference between the two lies in the fact that the former are thick sequences of different iron minerals, interbedded with chert and have been deposited in intracratonic basins, while the latter occurs in thick sequences, having oolitic character and deposited in localised basins.

14.2 MINERALOGY

The common iron minerals that are encountered in sedimentary rocks fall into four main catagories as; oxides, carbonates, silicates and sulphides. These mineral groups or facies include following minerals:

1. Oxides: Magnetite Fe_3O_4; Hematite Fe_2O_3
   Goethite $FeO{\cdot}OH$; Limonite $FeO{\cdot}OH{\cdot}nH_2O$
2. Carbonate: Siderite $FeCO_3$
4. Silicates: Chamosite $Fe_3Al_2Si_2O_{10}{\cdot}3H_2O$
   Glauconite $KMg(Fe{\cdot}Al)(SiO_3)_6{\cdot}3H_2O$
5. Sulphide: Pyrite FeS_2; Marcasite FeS_2

Many of these mineral groups are encountered within a single iron deposit. The different mineral groups also occur in rhythmically repeated layers in stratigraphic succession.Lateral changes in mineral groups within a single deposit have also been reported.

Occurrence of different mineral groups within a single deposit is a result of environmental factors,such as, Eh and pH (Fig.14.1) and also the availability of sulphide minerals. Stability

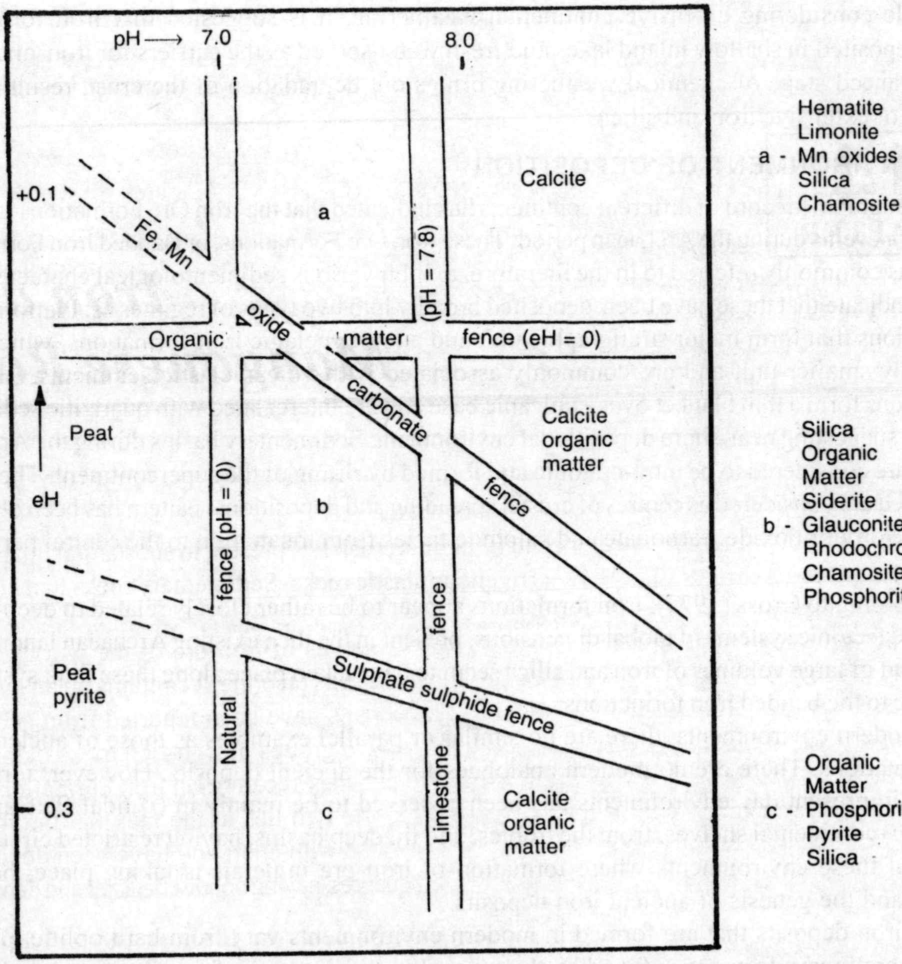

Fig. 14.1. Chemical classification of non–clastic sediments on the basis of Eh–pH.

relationships between these different mineral groups accordingly decipher the presence of a particular mineral group in iron-formation.

Sources of iron

Continental weathering and contemporaneous volcanicity are considered as two main sources of iron. Some of the Archaean iron-formations are associated with volcanic and iron is supposed to be derived from hydrothermal alterations and submarine weathering of basic extrusives, whereas, for most sedimentary iron-deposits, continental weathering is considered as the prominent source of iron. It is released due to intensive weathering of mafic minerals from igneous and metamorphic rocks, under humid tropical climatic conditions.

Due to close association of volcanics with Iron formations, volcanism is considered as a source of iron and also of silica. The then ocean water has been considered as the main source for both, i.e., Fe and SiO_2. The pH and oxidation potential of oceanic water are considered responsible for deposition of iron on extensive scale.

While considering extensive continental weathering, it is suggested that Iron formations were deposited in shallow inland lakes and fresh water served as the carriers for iron and silica. An advanced stage of chemical weathering brings out degradation of the crust, resulting in a release of extensive iron and silica.

14.3 ENVIRONMENT OF DEPOSITION

The stratigraphic record of different continents has indicated that the Iron Ore Formations is one of the major events during the Archaean period. These Iron Ore Formations, or Banded Iron Formations (BIF), as commonly referred to in the literature, exhibit various sedimentological characteristics, which indicate that these have been deposited broadly into two types of regimes as; Platform Iron-Formations that form major stratigraphic unit and another pelagic iron-formations, which form relatively smaller unit and are commonly associated with volcaniclastic sediments. The iron formations form a thin blanket over a volcanic base and are intercalated with quartzitic sediments, thereby suggesting near-shore depositional environment. Sedimentary basins during the Archaean period are considered to be intra-cratonic and formed by rifting of the supercontinent. The basins so formed are considered as centres of crustal spreading and depositional pattern has been observed from the order of oxide, carbonate and sulphide facies from the margin to the central part of the basin.

According to Gross (1973), iron formations appear to be rather closely related to deep-seated fault and tectonic systems of global dimensions, present in the then existing Archaean landmasses. Transport of large volumes of iron and silica seem to have taken place along these fault systems to give rise to the banded iron formations.

In modern environments, there are no similar or parallel examples as those of ancient iron-ore formations. There are no modern analogues for the ancient deposits. However, formation of iron in present day environments has been observed to be mainly in (*i*) tidal flats and bog lakes, (*ii*) continental shelves, from the tropics, and iii. deep basins, having restricted circulation. Study of these environments where formation of iron ore minerals is taking place, helps to understand the genesis of ancient iron deposits.

The iron deposits that are formed in modern environments vary from hard oolitic, pisolitic and concretionary forms to soft and earthy types. Iron minerals, so formed, are rather difficult to identify due to their amorphous or partly crystalline nature.

An important problem concerns with the transportation of iron to the basin of deposition, where it is separated from other materials to form iron mineral deposits. It has been observed that iron content of stream or river waters is sufficiently high to account for the formation of iron deposits. Ferric compounds are notably insoluble, while ferrous oxides and hydroxides generally accumulate in soils and laterite—as residual deposits. Ferrous salts, like ferrous bicarbonate are readily soluble, but being insoluble in the presence of oxygen, are prone to hydrolysis, which then get precipitated. More probably, most of the iron in modern day environments appear to be transported in the colloidal form of iron oxide It, therefore, enters into the clay mineral fraction. The iron thus transported gets released under reducing conditions to form the minerals like glauconite and chamosite.

Precipitation of iron in the form of ferrous carbonate is a function of prevailing condition in a local environment. Prevalance of reducing conditions locally is responsible in formation of ferrous sulphide. Iron compounds to be precipitated are, therefore, primarily a function of oxidation-reduction potential, and which explains the stability of different iron-bearing minerals more commonly than the acidity-alkalinity of the medium from which iron minerals precipitate. On the basis of oxidation-reduction potential, it has been observed that at the lowest potential only iron sulphide will form, while at a slightly higher potential ferrous carbonate is stable and ferric hydroxides are formed under fully oxidising conditions. It appears that magnetite formation takes place rather at a lower oxidising potential than other iron oxide minerals.

Accordingly, naturally occurring different mineral facies have been reported. Iron sulphides generally occur alone or in association with siderite. Iron silicate occurs in association with siderite and magnetite; while iron oxides occur alone or with iron silicates. Iron oxide minerals alongwith some silicates, like glauconite, exhibit features like granules or oolites and occur in association with quartz grains. Such an assemblage is an indication of turbulent environment at the time of formation of these iron minerals, and more so in an aerobic rather than in an anaerobic conditions. Attempts have been made to establish the relation between these different mineral facies and physico-chemical conditions under which they form. However though, no specific relation has been established, timé-equivalent lateral gradation from one mineral facies into another has been noted. These are found in oxide to chamosite, hematite to magnetite, to silicate and carbonate.

The stratigraphic record reveals that bulk of the banded iron formations have been formed during the Archaean period, and for their formation, controlling factor probably was the composition of ocean water, which was the major reserve of both iron and silica. Their source was rather volcanogenic and partly terrestrial. Volcanic source for the iron and silica cannot be ruled out due to close association of volcanic rocks with banded iron formation. According to Gross (1986), hydrothermal effusive processes seem to be the principal sources of metal as constituents in stratified iron formation. Earlier, Holland (1973) had suggested upwelling of deep ocean water as an alternative source of iron in banded iron formations. While upholding the view of Holland, Morris (1986) accounts for the deposition of banded iron formations from upwelling ocean currents, as chemical sediments in a platformal environment.

Thus, various theories account for the origin and environment of deposition of the banded iron formation, yet no theory seems to be complete in itself to satisfactorily explain the formation and deposition of BIF.

14.4 PRECAMBRIAN IRON FORMATIONS

The Iron formation during the Precambrian is one of the major events of the geologic time on all the Archaean schist belts of the world. These are being explored for rich deposits of iron ore, which they host. However, significant are the special conditions of formation and deposition of iron ore that prevailed on a global scale. It is also important to note that such conditions have not repeated later in geologic time. Precambrian iron formations, therefore, throw light on origin, palaeoclimatic conditions and tectonic environment that prevailed on all the then Archaean Nucleii.

More commonly, iron ore formation in literature is referred to as, Banded Magnetite Quartzite (BMQ) and Banded Hematite Quartzite (BHQ). Indian occurrences in Bihar and Orissa, have the stratigraphic nomenclature as, Iron-ore Series and Iron-ore Group However, the term, `Banded Iron Formation' (BIF) has found a wider acceptance.

The Precambrian Iron Formations have been classified into two types by Gross (1965), as, Algoma and Superior types. The Algoma type is mainly Archaean in age, while the Iron formations of Superior type are of Proterozoic age. The distinctive characteristics between the two can be summarised in the following paragraphs.

The Algoma type Banded Iron Formations are generally more than 2600 my. in age. However, few of Proterozoic and of Phanerozoic age have also been recorded. These formations have been deposited in oceanic tectonic basins, in marginal part associated with greenstone belts. These commonly occur as irregular, lenticular bodies within the Archaean basement rocks. Mafic volcanics and graywackes are generally associated with them. The sedimentary facies commonly encountered is oxide facies, with subordinate carbonate and sulphide facies. It is observed that sulphide and carbonate facies occur near the center, while oxide facies has developed at the margin of the basin.

In contrast to Algoma type, the BIF of Superior type are the characterisics of the stable continental shelves, deposited in shallow water environment and are found to be confined to intracratonic basins. These are of Proterozoic age and occur as laterally extensive bodies. The associated rocks are quartzite, conglomerate, dolomite and black shale. There is no direct evidence of association with volcanics, which are normally absent. In this type, oxide facies is abundant, while sulphide facies is nearly absent.

Apart from the Algoma and Superior types, there is another type of iron formation, which is confined mainly to granulite terrain. It is a poorely-banded magnetite-quartzite in a supracrustal sequence of quartzites, mica-schists, marbles, and para-amphibolites. This entire sequence is engulfed into voluminous tonalitic gneiss. The BIFs from Karnataka, Tamilnadu and Kerala from southern India are associated with the granulitic terrain.

The Iron Formations of India are found to have the characterisics of both the types, i.e., the Algoma and the Superior. The BIF from southern Karnataka are mainly of Algoma type, whereas those from Bihar and Orissa are classified as of Superior type.

There are two prominent events, of iron formations in India. The older one is greater than 3000 my. in age, and is from southern India, while the younger one is restricted to the schist belts, ranging in age between 2900-2600 my. and which have the similar characteristics as those of Algoma type. It is significant that extensive Middle to Late Proterozoic sedimentary basins are devoid of any iron formations of Superior type.

On the basis of the dominant iron-bearing mineral, iron-formations have been grouped into four distinct facies, as; oxide, carbonate, silicate and sulphides.

14.5 MANGANESE NODULES

One of the richest and economically significant and attractive authigenic metal rich sediments that occur on the ocean floor are manganese nodules. It has been observed that apart from manganese, nodules contain billions of tonns of metals, such as; nickel, copper, cobalt, iron and traces of other metals. Nickel is particularly important as there are few deposits and that also occur in copper rich nodules on the ocean floor. Cobalt is another significant metal and is found to occur with concentration of iron and lead. Study of the distribution pattern suggests that nodules with, nickel and copper are generally common in areas of siliceous oozes, whereas iron is mostly of hydrothermal origin.

A wide-spread occurrence of manganese rich sediments has been reported from different ocean floors. Manganese occurs as nodules, crusts or as thin coatings on rocks. Though, these are widely distributed on the ocean floor, they do not form in areas where rate of sedimentation is high and turbidites and hemipelagic sediments are deposited. Favourable areas for their formation are the areas characterised by low rates of sedimentation, i.e. less than 5 mm. per 1000 yrs. and where well-oxygenated ocean floor conditions prevail.

Manganese nodules exhibit concentric layered structure. It results due to changes in composition and mineralogical boundaries. Layers range in size and can be distinguished visually and also by examining under microscope. Such a layered structure results due to changes in the environment of formation, reflecting palaeogeochemical indicators, which are related to changes in ocean water chemistry during the growth of nodules. It has been noted that incorporation of biogenic sediment particle in the nodule matrix takes place during periods of high pelagic sedimentation or low metal-oxide accretion.

Precipitation of MnO_2 in the form of nodule needs catalytic effect of a precipitation surface. It, therefore, requires a nucleus or rock surface to start with. There are two principal minerals, namely; birnessite and todorokite. These minerals differ mainly in the degree of oxidation and hydration, which is controlled by redox potential of the geochemical environment of deposition.

Changes in nodule morphology and chemical composition have also been reported from different ocean water regimes. Studies have indicated that in the upper regime, chemical reactions take place between nodule and sea water; whereas, in the lower regime, reactions take place between sea water, the nodule and sediment. This results in change in chemical composition of nodule. It is found that Mn and Cu greatly concentrate in the lower regime, and concentration of Fe and Co is in the upper regime.

Rate of growth of manganese nodules has been estimated by radiometric methods and it has been found that growth rate varies between 1 and 4 mm./10^6 yrs. or from 0.2 to 1.0 mgm. cm^2/1000 yrs.

Occurrence of manganese nodules on the Carlsberg Ridge has already been reported. The Carlsberg Ridge shows a rugged topography of ridges and valleys. These are covered by sediments with a maximum thickness of about 60 m. on the southwestern flank of the ridge. These sediments are white to dark brown calcareous oozes. It has been observed that foraminiferal oozes commonly occur on the topographic highs above 3500 m., while siliceous oozes are encountered in valleys, and surficial deposits of manganese nodules have been reported on the slopes of the ridges.

These manganes nodules are of seven different types and are composed of Fe, Ni, Cu, Co and Zn, apart from Mn. It has also been observed that with increasing diameter of the nodules, there is a corresponding increase in Mn, Cu and Co content and decrease in Fe and Zn content (Valsangkar, *et al.* 1988); while there is no significant variation in Ni content. Manganese nodules from Central Pacific Ocean show an increase in Mn, Ni and Cu with a corresponding increase in minimum diameter of the nodules. It has also been observed that rough surface nodules are generally richer in Mn, Ni and Cu than the smooth surface nodules.

EPILOGUE

Epilogue

The book 'Applied Sedimentology' contains Fourteen chapters. Though the title of the book indicates 'applied' aspects, it cannot bypass essential and fundamental theoritical aspects of sedimentology. The first chapter, 'Introduction', narrates in brief, relationship of 'sedimentology' with other branches of earth science and also how its premises are based on the subjects like; Physics, Chemistry, Biology and the like.

Second chapter on 'Field Methods' helps students and young researchers in sedimentology, in what way to proceed in the field to collect representative samples. It also brings out limitations that are encountered during the field investigations. Sedimentological data to be obtained in laboratory is based on 'samples' and hence is the importance of samples and methods of sampling in the field.

The third chapter relates to the modes of formation of sediments and describe the stages of the sedimentary cycle. Origin of sediments naturally follows 'sedimentary cycle'. Detailed textural analysis has been dealt with accordingly, elaborating the statistical concept and statistical analysis, and interpretation of textural data therefrom. Recently introduced concept of 'fractal' in textural analysis has also been included. The chapter is oriented mainly on the lines of research methodology.

Fundamentals of sediment movement and their significance in sedimentology, particularly in the formation of sedimentary structures have been indicated in the fifth chapter. Theory of sediment movement also finds application in the field of Petroleum Geology. Sedimentary structures, therefore, alongwith their mode of origin, classification, morphology and analytical aspects have been detailed in the Sixth chapter. Various primary and sedimentary structures have been dealt with, including their field characters.

Classification of sedimentary rocks follows in the Seventh Chapter. Various schemes of classification, beginning from the older one till those which are being followed in present day teaching and research, have been presented in different tabular and diagrammatic forms. Classification of sedimentary rocks according to their mode of formation has been discussed initially, followed by the detailed classification of each group.

Origin and formation of various sedimentary rocks take place in a variety of depositional sedimentary environments. These rocks have, therefore, specific characters of their own. This has given rise to the concept of 'Facies'. 'Sedimentary Facies' or environments of deposition bear close relationship with tectonics in the global context, which was identified earlier with the 'Geosynclinal theory'. The concept of 'Geosyncline' is now being discussed in the light of the theory of 'Plate Tectonics' and sedimentary basins have accordingly been identified and classified. These aspects have been discussed at length in the chapter, 'Sedimentary Facies and Products'.

Of the variety of sedimentary rocks, sandstone and carbonate rocks form the major bulk. These two varieties, therefore, have been described and dealt with in detail, referring to the aspects of mineral composition, cementation and diagenesis. These rocks have been dealt with in the chapters, as 'Sandstone' and 'Carbonate Rocks'.

Minor sedimentary rocks include Evaporites, Chert, Phosphate and Iron and Manganese rocks. These minor sedimentary rocks have been described with their relevant and important aspects; such as; theories of their origin, mineralogical and textural aspects, earlier deposits with their modern analogs.

Extensive literature has been published and also is being published by way of research papers and in the form of books, in the field of sedimentology. Common experience reveals that a university/college teacher and researcher in the field of sedimentology have to scan the wealth of literature to get the relevant information. This book is an attempt to encompass all such information in the concise form, which, it is hoped will prove useful to them. A common reader in Geology may find the book, as giving an insight in the field of sedimentology alongwith its applied aspects.

Bibliography

Adams, J. E. and Rhodes, M. L., 1960. Dolomitisation by seepage reflux., *A. A. P. G., Bull.*, **44**: 1912-1920.

Allen, T., 1968. *Particle Size Measurements*. Chapman and Hall, London, pp. 248.

Atterberg, A., 1905. Die rationelle klassification der Sande and Kiese., *Chem. Zeitschr.*. **29**: 195-198.

Badiozamani, K., 1973. The Dorag dolomitisation model—application to the Middle Ordovician of Wisconsin. *Jour. Sed. Pet.* **46**: 965-984.

Bagnold, R. A., 1973. The nature of saltation and of bed-load transport in water., *Roy. Soc. London, Proc., Ser. A*, **332**; 473-504.

Balakrishnan Nair, T. M., Ramaswamy, V., Parthiban, G. and Shankar, R., 1999. Factors Controlling Vertical Fluxes of Particles in the Arabian Sea. *Geol. Soc. Ind., Jour*. **54**: 369-378.

Belderson, R. H. and Stride, A. H., 1966. Tidal current fashioning of a basal bed. *Marine Geol.*, **4**: 237-257.

Blatt, H., 1972. *Sedimentary Petrology* (2nd ed.). W. H. Freeman and Co., New York, pp. 514.

... "..., Middleton, G. and Murray, R., 1980. *Origin of Sedimentary Rocks* (2nd ed.). Prentice Hall Inc., New Jersey, pp. 782.

Bouma, A. H., 1962. *Sedimentology of Some Flysch Deposits*. Elsevier, Amsterdam, pp. 168.

Brenchley, P. J. and Williams, B. P .J., (ed.) 1985. Sedimentology: Recent Developments and Applied Aspects. Blackwell Sci. Publ., pp. 342.

Collinson, J. D. and Thompson, D. B., 1984. *Sedimentary Structures* (2nd ed.). George Allen and Unwin, London, pp. 194.

Davis, R. A. 1992. *Depositional Systems: An Introduction to Sedimentology and Stratigraphy*. (2nd ed.), Prentice Hall, New Jersey, pp. 604.

Dickinson, W. R., 1970. Interpreting detrital modes of graywacke and arkose. *Jour. Sed. Pet.*, **40**: 695-707.

Dickinson, W. R. and Suczek, C. A., 1979. Plate tectonics and sandstone compositions., *A.A.P.G., Bull.* **63**: 2164-2182.

Dunham, R. J., 1962. Classification of carbonate rocks according to depositional textures. *A.A.P.G., Mem.* **1**: 108-121.

Embry, A. F. and Klovan, J. E., 1972. A Late Devonian reef tract on north-eastern Banks Island, Northwest Territories. *Can. Petrol. Geol., Bull.*, **19**: 730-781.

Fairbridge, R. W. and Finkl, C. W., 1988. I.G.C.P. - 274, Coastal evolution in the Quarternary-Note. Inaug. Meet., The Netherlands, 1-4.

Folk, R. L., 1955. Student operator error in determination of roundness, sphericity and grain size. *Jour. Sed. Pet.* **25**: 297-301.

Folk, R. L., 1959. Practical petrographic classification of limestones. *A.A.P.G., Bull.* **43**: 1-38.

..."..., 1962. Spectral subdivision of limestone types. *A.A.P.G., Mem.* **1**: 62 - 84.

...''..., 1965. Some aspects of recrystallisation in ancient limestones.In : Dolomitisation and Limestone Diagenesis. *In:* Pray, L. C. and Murray, R. C. (eds.). *Publ., Soc. Econ. Palaent. and Min.* **13**: Tulsa. pp.14-48.

...''..., 1966. A review of grain-size parameters. *Sedimentology*. **6:** 73 - 93.

...''..., 1968, *Petrology of Sedimentary Rocks*. Hemphill. pp. 170.

...''..., 1974. *Petrology of Sedimentary Rocks*. Hemphill Publ. Co., Austin, Texas, pp. 159.

...''... and Ward, W. C., 1957. Brazos river bar: A study in the significance of grain-size parameters. *Jour. Sed. Pet.* **27:** 3-27.

...''... and Land, L. S., 1975. Mg/Ca ratio in salinity: Two controls over crystallisation of dolomite. *A.A.P.G., Bull.* **59:** 60-68.

Friedman, G. M., 1962. On sorting, sorting coefficients, and the log normality of the grain-size distribution of sandstones. *Jour. Geol.*, **70:** 737-753.

...''... and Sanders, J. E., 1978. *Principles of Sedimentology*. John Willey, New York, pp. 792.

...''..., ...''... and Kopaska-Merkel, D. C., 1992. *Principles of Sedimentary Deposits: Stratigraphy and Sedimentology*. Macmillan, Publ. Co., New York, pp. 717.

Gadewar, A. J., 1989. Sedimentological studies of the Upper Gondwana Sandstones, Belkher-Bairamghat area, Amravati-Betul Dists., Maharashtra and Madhya Pradesh., Unpubl. M.Phil. thesis, submitted to University of Poona, Pune, pp. 79.

Geological Society of India, 1992. *Thar Desert in Rajasthan. In:* Singhvi, A. K. and Kar A. (eds). Spl. Publ., pp. 191.

...''..., 1986. Banded Iron Formations of India. *Spl. Issue*, **28** (2 & 3): 71-276.

Gokhale, A. P., 1981. Sedimentological studies of the Kaladgi sandstones from the Jamkhandi area, Bijapur Dist., Karnataka., Unpubl. M.Phil. thesis, submitted to University of Poona, Pune, pp. 93.

Grabau, A. W., 1904. On the classification of sedimentary rocks. *Amer. Geol.* **33:** 228-247.

Griffith, J. C., 1961. Measurement of the properties of sediments. *Jour. Geol.* **69:** 487-498.

...''..., 1967. *Scientific Methods in Analysis of Sediments*. McGraw Hill Book Co., New York.

Gross, G. A., 1965. The geology of iron deposits in Canada: General geology and evaluation of Iron Deposits. *Geol. Surv. Canada, Econ. Geol. Ser. Rep.* **22**(1): 181.

...''..., 1973. The depositional environments of the principal types of Precambrian Iron Formations. *In: Genesis of Precambrian Iron and Manganese Deposits. Earth Sciences*, UNESCO, Paris, **9**: 15-21.

...''..., 1986. The metallogenetic significance of Iron-Formation and related stratafer rocks. *Geol. Soc. Ind. Jour.*, Spl. Issue, Banded Iron Formations of India, **28**(2&3): 92-108.

Hakanson, L. and Jansson, M., 1983. *Principles of Lake Sedimentology*. Springer Verlag, Berlin. pp. 316.

Hutchinson, G. E., 1957. *A Treatise on Limnology*. John Wiley and Sons, New York. p. 1.

Ingram, R. L., 1954. Terminology for the thickness of stratification and parting units in sedimentary rocks. *Geol. Soc. Am., Bull.* **65**: 937-938.

Inman, D. L., 1952. Measures for describing the size distribution of sediments. *Jour. Sed. Pet.* **22**: 125-145.

Joshi, A. V., 1993. Origin of chert in Lameta Limestone (Maestrichtian) of Gujarat State. *Bull. Pure and Appl. Sci.* **12**F(1&2): 1-8.

Kennet, J. P., 1982. *Marine Geology*. Prentice Hall, New York, pp. 813.

Klein, G. deV., 1963. Analysis and review of sandstone classification in the North American geological literature. *Geol. Soc. Am. Bull.* **74**: 555-576.

Kleinspehn, K. L. and Paola, C. (ed.), 1988. *New Perspectives in Basin Analysis (Frontiers in Sedimentary Geology)*. Springer Verlag Inc., New York. pp. 453.

Krumbein, W. C., 1934. Size frequency distribution of sediments. *Jour. Sed. Pet.* **4**: 65-77.

...''... and Sloss, L. L., 1963. *Stratigraphy and Sedimentation* (2nd ed). W. H. Freeman and Co. San Fransisco. pp. 660.

Krynine, P. D., 1948. The megascopic study and field classification of sedimentary rocks. *Jour. Geol.* **56**: 130-165.

Kuenen, Ph. H., 1953. Significant features of graded bedding. *A.A.P.G., Bull.* **37**: 1044-1066.

Kuenen, Ph. H. and Menard, H. W., 1952. Turbidity currents, graded and non-graded deposits. *Jour. Sed. Pet.* **22**: 83-96.

..."5..., and Migliorini, C. I., 1950. Turbidity currents as a cause of graded bedding. *Jour. Geol.* **58**: 91-127.

Kulkarni, S. A., 1993. Textural and Mineralogical studies of the Coastal Sediments between Karli Creek to Talavda Creek, Sindhudurg Dist., Maharashtra, India., Unpubl. Ph.D. thesis, submitted to Shivaji University, Kolhapur, pp. 322.

Langbein, W. B. and Schumm, S. A., 1958. Yield of sediment in relation to mean annual precipitation. *Am. Geophy. Union, Trans.* **39**: 1076-1085.

Mali, P. D., 1994. Sedimentological studies of the Kaladgi Group of rocks around Bagalkot, Dist., Bijapur, Karnataka., Unpubl. Ph.D. thesis, submitted to Shivaji University, Kolhapur. pp. 132.

Mandlebrot, B. B., 1977. *A Fractal Geometry of Nature*. Freeman, San Franscisco.

Miall, A. D., 1984, *Priciples of Sedimentary Basin Analysis*. Springer Verlag, New York. pp. 490.

..."..., 1997. *The Geology of Stratigraphic Sequences*. Springer, New York. pp. 433.

Moore, R. C., 1949. Meaning of facies. *Geol. Soc. Am., Mem.* **39**: 1-34.

Morris, R. C., 1986. The cycling redox state of iron in the genesis of Banded Iron Formations and their associated enrichment iron ores. *Geol. Soc. Ind., Jour.*, Spl. issue, Banded Iron Formations of India. **28**(2 & 3): 227-236.

Okada, H., 1971. Classification of sandstone: Analysis and Proposal. *Jour. Geol.* **79**: 509-525.

Orford, J. D. M and Whalley, W. B., 1983. The use of fractal dimension to quantify the morphology of irregular-shaped particles. *Sedimentology*. **30**: 655-668.

Otto, G. H., 1938. The sedimentation unit and its use in field sampling. *Jour. Geol.* **46**: 569-582.

Pandit, S. J., 1992. Geological studies of the Coastal tract around Ratnagiri, Ratnagiri Dist., Maharashtra. Unpubl. Ph.D. thesis, submitted to Shivaji University, Kolhapur. pp. 313.

Pandian, R. S., 1988. Textural studies of the Coastal Quaternary Sediments along the West Coast between Agashi Creek and Bassein Creek, Dist. Thane, Maharashtra, India., Unpubl. Ph.D. thesis, submitted to University of Poona, Pune. pp. 138.

Pawar, J. B., 1988. Sedimentological studies of the Kargil Formation around Kargil, Dist., Ladakh, Jammu and Kashmir State, India., Unpubl. M.Phil. thesis, submitted to University of Poona, Pune. pp. 79.

Pettijohn, F. J., 1948. A preface to the classification of sedimentary rocks. *Jour. Geol.*, **56**: 112-118.

..."..., 1957. *Sedimentary Rocks*.(2nd ed.,) Harper and Bros., New York.

..."..., 1984. *Sedimentary Rocks* (3rd ed.)., CBS Publ. and Distributors, Delhi. pp. 628.

..."..., Potter, P. E. and Siever, R., 1973. *Sand and Sandstones*. Springer-Verlag, Berlin. pp. 618.

Powers, M. C., 1953. A new roundness sçale for sedimentary particles. *Jour. Sed. Pet.* **23**: 117-119.

Reeves, C. C., jr., 1968. Introduction to Palaeolimnology. *In*: *Developments in Sedimentology*., V. 11, Elsevier, Amsterdam. pp. 228.

Reineck, H. E., 1971. Der kustensand., *Natur. Mus.* **101**: 45-60.

..."..., and Singh, I. B., 1980. *Depositional Sedimentary Environments*. Springer-Verlag, Berlin, pp. 439.

Ruxton, B. P. and McDougall, I., 1967. Denudation rates in north-east Papua from Potassium-Argon dating of lavas. *Am. Jour. Sci.*, **265**: 545-561.

Schrock, R. R., 1946. Classification of sedimentary rocks. *Geol. Soc. Am., Bull.* **57**: 1231.

..."..., 1948. Classification of sedimentary rocks. *Jour. Geol.* **56**: 118-120.

Schumm, S. A., 1968. Speculation concerning palaeohydrologic controls of terrestrial sedimentation. *Geol. Soc. Am.* 79: 1573-1586.

Selley, R. C., 1976. *An Introduction to Sedimentology*. Academic Press, London. pp. 408.

Sneed, E. D. and Folk, R. L., 1958. Pebbles in the lower Colorado River, Texas-A study in particle morphogenesis. *Jour. Geol.*, **66**: 114-150.

Seilacher, A., 1964. Biogenic sedimentary structures. *In*: *Approaches to Palaeoecology*, Imbrie, J. and Newell, N. D. (eds)., John Wiley and Sons, New York. pp. 296-315.

Sukhtankar, R. K., 1975. Geology of the area around Muddebihal, Dist. Bijapur, Karnataka State, India., Unpubl. Ph.D. thesis, submitted to University of Poona, Pune. pp. 209.

Suryanarayan, K. and Krishna Rao, G., 1996. Geochemical changes due to weathering under Different Environmental Conditions in parts of Eastern Ghats. *Geol. Soc. Ind. Jour.* **48**: 427-437.

Trask, P. D. (ed)., 1939. Recent Marine Sediments. A.A.P.G., Tulsa, Oklahoma. pp. 736.

Tucker, M. E., 1981. *Sedimentary Petrology: An Introduction*. ELBS, Blackwell Scientific Publ., Oxford. pp. 252.

Twenhofel, W. H., 1932. *Treatise on Sedimentation*. 2nd ed. The Williams and Wilkins Co., Baltimore pp. 926.

Udden. J. A., 1898. Mechanical composition of wind deposits. *Augustana Lib. Publ.* **1**: 69.

Wadell, H., 1932. Volume, shape and roundness of rock particles. *Jour. Geol.* **40**: 443-451.

Walker, R. G., 1976. Facies Models 3: Sandy Fluvial Systems. *Geosci. Canada*. **3**: 101-109.

Wentworth, C. K., 1922. A scale of grade and class terms for clastic sediments. *Jour. Geol.* **30**: 377-392.